NASA SP-482

THE IMPACT OF SCIENCE ON SOCIETY

James Burke
Jules Bergman
Isaac Asimov

I0473821

Prepared by Langley Research Center

NASA Scientific and Technical Information Branch   1985
National Aeronautics and Space Administration
Washington, DC

Library of Congress Cataloging in Publication Data

Burke, James, 1936–
   The impact of science on society.

   (NASA SP ; 482)
   Series of lectures given at a public lecture series sponsored by NASA
and the College of William and Mary in 1983.
   1. Science—Social aspects—Addresses, essays, lectures. I. Bergman,
Jules. II. Asimov, Isaac, 1920–   . III. United States. National
Aeronautics and Space Administration. IV. College of William and
Mary. V. Title. VI. Series.
Q175.55.B88   1985   303.4'83   84-14159

For sale by the Superintendent of Documents, U.S. Government Printing Office
Washington, D.C. 20402

Foreword

Science and technology have had a major impact on society, and their impact is growing. By drastically changing our means of communication, the way we work, our housing, clothes, and food, our methods of transportation, and, indeed, even the length and quality of life itself, science has generated changes in the moral values and basic philosophies of mankind.

Beginning with the plow, science has changed how we live and what we believe. By making life easier, science has given man the chance to pursue societal concerns such as ethics, aesthetics, education, and justice; to create cultures; and to improve human conditions. But it has also placed us in the unique position of being able to destroy ourselves.

To celebrate the 25th anniversary of the National Aeronautics and Space Administration (NASA) in 1983, NASA and The College of William and Mary jointly sponsored a series of public lectures on the impact of science on society. These lectures were delivered by British historian James Burke, ABC TV science editor and reporter Jules Bergman, and scientist and science fiction writer Dr. Isaac Asimov. These authorities covered the impact of science on society from the time of man's first significant scientific invention to that of expected future scientific advances. The papers are edited transcripts of these speeches. Since the talks were generally given extemporaneously, the papers are necessarily informal and may, therefore, differ in style from the authors' more formal works.

As the included audience questions illustrate, the topic raises far-reaching issues and concerns serious aspects of our lives and future.

Donald P. Hearth
Former Director
NASA Langley Research Center

Contents

The Legacy of Science

James Burke

James Burke

For more than a decade, James Burke has been one of the British Broadcasting Corporation's outstanding television writers, hosts, and producers. Born in Northern Ireland and educated at Oxford University, Burke spent 5 years in Italy teaching at the Universities of Bologna and Urbino and directing the English Schools in Bologna and Rome. He made his television debut in 1965 as a reporter for Granada Television's Rome Bureau.

Burke's impressive following in the British Isles dates back to 1966, when he joined the BBC's weekly science show, *Tomorrow's World.* As the chief BBC correspondent for all Apollo space flights, Burke won critical acclaim for his interpretation of the US space program to an audience of over 12 million people. During this time he developed and presented a variety of documentaries, and in 1972 he became the host of his own weekly prime-time science series, *The Burke Special.* The programs earned for Burke a Royal Television Society Silver Medal in 1972 and a Gold Medal in 1973. In 1975-1976, Burke co-authored and co-hosted *The Inventing of America,* an NBC/BBC joint production for the US Bicentennial.

Burke's 10-part television series *Connections,* which aired in 1979, attracted one of the largest followings ever for a Public Broadcasting Station documentary series, and the companion book was a bestseller in both the UK and the US. The series, which took a year of research and another year to film at more than 100 locations in 22 countries, surveyed the history of technology and social change by tracing the evolution of eight major modern inventions: The atom bomb, telecommunications, computers, production lines, jet aircraft, plastics, rocketry, and television. In 1980 Burke wrote and presented *Burke: The Real Thing,* a BBC six-part series on reality and human perception. He is a regular contributor to such major magazines as *Vogue, The Atlantic Monthly, Harpers, New York Magazine,* and *New Scientist.*

The Legacy of Science

Change is one of mankind's most mysterious creations. The factors that operate to cause it came into play when man produced his first tool. With it he changed the world forever, and bound himself to the artifacts he would create in order, always, to make tomorrow better than today. But how does change operate? What triggers a new invention, a different philosophy, an altered society? The interactive network of man's activities links the strangest, most disparate elements, bringing together the most unlikely combinations in unexpected ways to create a new world.

Is there a pattern to change in different times and separate places in our history? Can change be forecast? How does society live with perpetual innovation that, in changing the shape of its environment, also transforms its attitudes, morals, values? If the prime effect of change is more change, is there a limit beyond which we will not be able to go without anarchy, or have we adaptive abilities, as yet only minimally activated, which will make of our future a place very different from anything we have ever experienced before?

Somebody once apparently said to the philosopher Wittgenstein, "What a bunch of no-knows we medieval Europeans must have been, back in the days before Copernicus, to have looked up at the sky and thought that what we saw up there was the Sun going round the Earth, when, as everybody knows, the Earth goes round the Sun, and it doesn't take too many brains to understand that!" Wittgenstein replied, "Yes, but I wonder what it would have looked like if the Sun *had* been going round the Earth." The point is that it would, of course, have looked exactly the same. What he was saying was that you see what you want to see.

Consider also the medieval Londoner or eighteenth-century American who, when asked what he thought of the prospect that

3

one day everybody would have his own individual form of personal transportation, laughed at the idea of the metropolis at a standstill when the streets became, as they surely would, 14 feet deep in horse manure. The concept of any other form of transportation was outside his context.

I started with these two stories because they illustrate what I want to talk about today. If you look back at the cultural history of the West (and I do this on the premise that you only know where you're going if you know where you've been, and that those who are not prepared to learn the lessons of history are condemned to repeat it), the most important thing about the process of change and forecasting change at any one time hasn't been a matter of understanding the inner workings of the new gismos that mankind creates to make tomorrow better than today. In many cases, it has been the awareness that change was even happening at all, the understanding that the solid base from which prediction was being made might be about as solid as quicksand.

Even the most apparently immutable system or structure may be experiencing change even as you look at it. By change I mean, of course, not just change in the sense of bigger and better models and new ways of doing the same things you were doing before, but qualitative change in the structure and behavior of the society in which you live. I mean new philosophies as well as new gadgets.

I'm not saying that the appreciation of change is easy; far from it, particularly today. As one of your more respectable social forecasters said recently, "If you understand something today, that means it must by definition already be obsolete." Our general relationship with the present accelerating rate of change reminds me of the postcard from the patient on holiday to his psychiatrist at home: "Having a wonderful time. Why?"

The difficulty in recognizing change even when you fall into it, and the consequent off-the-cuff variety of forecasting that prevails, is, like most things (as I've just said), a matter of context. If you're looking to assess the future performance of an artifact or a human system within the envelope delineated by the factors involved, then what is the envelope, and how much of it are you aware of when you yourself are in the envelope? I'd like to start off by looking at that first.

Let me look at the envelope from a very basic point of view, that of the neurophysiology of raw perception itself. Forgive me if it's a bit oversimple. Take me—on the back of your retina I'm upside down, focused at the center but fuzzy at the edges, two-dimensional, a barrage of photons releasing rhodopsin and triggering neural impulses along the visual nerve. At the same time, the pressure wave I'm setting up right now with all this talk is causing little hairs inside the cochlea, in your inner ear, to shake around and send neural impulses into your brain. At no level am I aurally or visually more than a complicated version of the same neural impulse you'd get if you scratched yourself.

So what is it that makes all that mush *me*? After all, you've never met me before, and yet here I am, identified by you with absolute certainty as a human being, male, standing more or less upright, talking, and doing all the other things you've already recognized. What accomplishes that recognition job for you is your cognitive model. This is the construct, both experiential and genetic in origin, that you use to check up on all the separate bits of me and everything else you experience, mentally and physically, for identification purposes. You are using a recognition system made up of dedicated cells, each one firing in reaction to the one highly specific bit of detail it's built to react to.

Interestingly, it may be that the genetic component in that model is greater than was once thought. Some work going on here and in the UK shows that on the back of the human embryo, very early on in its development, is the neural plate, which contains the nerves that will eventually expand to form the spinal cord and the brain—the nervous system, in fact. Apparently, embryonic development involves millions of those little baby neurons growing and traveling (thanks to some kind of genetic zip coding) to an exact position in three-dimensional space in the final brain, in order for you to function at all when you're born and to lay down the basic matrix of interrelating neurons which will be the neurophysiological infrastructure of your personality and of the way you think.

I labor the point to show that the cognitive model isn't merely some psychologist's fantasy. It has a physical existence as a recognition system acting in an individual way to determine what

5

reality is or is not for the organism. Anything that doesn't get recognized in the most basic sense by the model is, naturally enough, rejected as meaningless. Of course, meaning is defined by your neurophysiological construct, not anybody else's.

Perception is directed and controlled by this cognitive model, both at the individual level and at the level of whole societies. Both kinds of models are very idiosyncratic. The Italian model has a sign like a wave, meaning, "Come here." Greek girls cause problems for non-Greek boys by saying "No" with a nod, not a shake, of their head. In New Zealand you can do one kind of V-sign but never the other. Americans look posh when they look neat; Europeans look posh when they look as if they've just come through a hedge backwards. A very fine linguistic example of model difference lies in the way the Irish and the English express themselves. Where the British will say that a situation is desperate but not hopeless, the Irish will say the situation is hopeless but not desperate.

The cognitive model, then, is what sets the rules, defines the structures, bestows meaning, sets up the ethics, values, beliefs, knowledge—everything that permits the user to function as a sentient organism, because it provides a perceived reality. It tells you which way is up, if you like. The model, then, determines what the universe is. As Wittgenstein said, "You see what you want to see."

Let me give you a good example of that axiom from history. Back in the twelfth century, when we were looking up at a Sun going round the Earth because Aristotle and the Church said that's the way it was, we were also looking up at what we thought was a perfect and unchanging universe, since if it had been created on day one of creation by the Deity, it had to be perfect. If it were perfect, there would be no change up there to see, so we didn't look up much. At the same time, however, the Chinese were busy getting a crick in the neck from doing just that—observing what was going on up there and logging everything that moved. They became expert astronomers centuries before we did, not because they were clever and we were dummies but because there was nothing in their model, as there was in ours, to stop them from seeing changes up there. We saw no change up in the sky because we thought there was none to see. Comets and supernovae were thought of as warnings from God!

The cognitive model—or to give it a better name, the paradigm—controls all decisions. If you believe the cosmos is made up of omelette, you build instruments specifically designed to find traces of intergalactic yolk. In that paradigm you reject phenomena like pulsars and black holes as paranormal garbage. In an omelette cosmos, the beginning of the universe becomes a chicken and egg problem, doesn't it?

Now, this definition of terms (like omelette universe) happens all the time. The reason that we today refer to electricity in terms of current is because in the eighteenth century people like Ben Franklin thought it was a fluid and tied all their experiments to that so-called fact. At the same time, in the eighteenth century, they thought all disease was caused by bad smells. Malaria is *mal aria*—bad air. We laugh at that, but for them, as for everybody at any time, the contemporary view of things is always the horse's mouth.

Every paradigm, at every time, in every place, is internally valid. By definition it has to be, for the organism or the group to function. Everybody has to have some version of reality—"the way things are for them," their definition of which way is up. This is perfectly valid at the time. All you can logically say about a guy who thinks he's a poached egg is that he's in the minority.

But then, if every paradigm is valid—set in philosophical concrete—what's the point in going through all that confusion when change happens? Metaphysically speaking, no one paradigm is innately any better than any other. A universe that began at 9 a.m. on October 10, 4004 B.C. (which was official back in the seventeenth century) is intrinsically no less valuable for those who live by a belief in it than is our present uncertain universe, perhaps built like a yo-yo, forever destroying and remaking itself in never-ending big bangs. Each of the cosmological theories has, at different times, found totally ironclad evidence to support it.

So, given that every paradigm in every place at every time has had epistemological reasons for being the only right one thus far, why does the boat get rocked time and again, why are waves made, why does change happen, when everything is fine as it is? Even if there are a few wrinkles in an otherwise fully adequate explanation of the universe, we try very hard to get around them.

For instance, something that bothered people back in the eighth century B.C. was the way the planets appear to go backwards

7

from time to time. According to Aristotle, who worked out a general theory in the fourth century B.C., the sky was made up of eight concentric crystalline spheres, each one carrying a planet and the outermost one carrying the stars, which were supposed to rotate east to west. This circular motion, being heavenly, was perfect—except for these planets going what looked like backwards from time to time. Well, that particular little inconvenience was solved by putting each planet in a little epicycle, or mini-orbit, spinning round and round while still remaining attached to its own individual main sphere, which still rotated east-west like it was supposed to. That way, the planets only appeared to go backwards sometimes.

The real explanation, that the Earth was moving as well and that this caused the appearance of retrograde motion, was unacceptable within the cosmological paradigm that was still operative in the Renaissance. It was unacceptable because it would have had philosophical and theological implications that were too hot to handle. The Bible would have been seen to be wrong, for example, because it said the Earth didn't move. So epicycles fit the bill, and kept things the way they were supposed to be. However, you had to have over 70 of them, and even then they didn't work absolutely perfectly.

"Saving the appearances at all costs" in that way is generally how we react to little inconsistencies in our paradigm. If your paradigm is the rock of ages for you—and it always is—then you let go of it only with immense reluctance. We are, paradigmatically speaking, extremely conservative.

Look at how often change is fought in history. Here's an example that always tickles me. The chain of events back in the twelfth century that set Europe going economically after the Dark Ages was essentially the textile revolution. A new loom came in from Arab Spain. It had foot pedals, which left the weaver's hands free to weave faster and make more cloth cheaper. The Dutch weavers smashed the thing up because it would have put people out of work. (That was a new idea in the twelfth century.)

A generation later, when the dust had settled, in came the spinning wheel from left field—a total surprise from China. It made thread very much faster than before. When the wheel and the loom were put together, the production of cloth skyrocketed. So there were more riots, because the cloth was linen, which was made from a plant and was cheaper than feeding sheep and making wool, so the rioters were sheep farmers.

But soon everybody was wearing linen, because it was cheap, and throwing it away when they wore holes in it. So there was this giant pile of linen rag lying around fourteenth-century Europe. The price of paper dropped like a stone, because linen rag paper was the best you could make. There were more riots—sheep farmers again, because parchment was sheepskin, and it had become too expensive to use.

So here was enough paper around to put on the walls, and the scribes were going like gangbusters and pretty soon they were on strike for higher wages because it was a seller's market. Everybody wanted their paperwork done because the Black Death was just over and everybody was inheriting like crazy. There just wasn't enough writing ability to go around, until Gutenberg came along in 1450 with the printing press.

Now this was something the Church wanted like a hole in the head, because it would encourage free thinking—until they realized that you could print indulgences with it. People bought the indulgences, because when they did that they got remission of some sins. With all the demand for instant salvation that followed, the Church made a million—money to build the Vatican, pay Michelangelo's bill, and generally get involved in prestigious projects that made certain German clerics really mad at this consumerist, money-making approach to religion. One of these Germanic chaps nailed up his criticisms, and there was the Reformation.

It's a little oversimplified, maybe, but you get my drift. People in general would rather fight than switch. So, to repeat myself, if the paradigm fits and people resist innovation, why does change happen at all? Well, let me give you some examples of the mechanisms that operate to produce change, and you'll see why it isn't that simple.

To begin with, often you just don't know change is coming. Even if you're personally involved, you may be looking the wrong way at the time, like young William Perkin of London in 1856. Around then, everybody was looking for benzene rings and chemistry was the flavor of the month, and Perkin, a chemist, was trying to be the young science hero who would save the great British empire by discovering the way to make artificial quinine chemically. You see, our administration and army chaps were dropping like flies out in the Far Eastern colonies because of malaria, and artificial quinine would have fixed things up right. Besides that, we were having to buy natural quinine from the Dutch in Java, and they charged an outrageous price for it. So that great motivator, money, was also at work. Well, after a bit Perkin came up with some interesting sludge, but one thing it wasn't was artificial quinine, so he threw it down the sink, and discovered that he had invented the world's first aniline dye. Made a million.

Sometimes, though, you may be looking in the right direction, but you don't see what's happening. In 1778, just after you people had gone off on your own and left us with no more South Carolina pitch to put on the bottom of our ships to protect them from rot, the rather seedy ninth Earl of Dundonald in Scotland thought up a plan to recoup the family fortunes by getting tar out of the coal from a couple of mines on his land. This tar would replace the pitch and make Dundonald a rich man. Unfortunately, the British government had already shifted to copper-bottoming its ships, so Dundonald's coal-heating kiln, where he made the tar, was useless, and so were the vapors he had been watching coming out of the kiln. In fact, he'd even been lighting them and generally playing around, shooting flames out of a tube. He happened to mention this to his friend James Watt, and three years later, Watt's sidekick "invented" coal gas. Dundonald died in poverty.

However, even when you get what you're looking for and you know you've got it, things can go haywire. Take Benjamin Huntsman, clockmaker, looking for a better clock spring in 1740 because pendulum clocks were no good at sea and you needed a clock to work out longitude, and in an era of great maritime expansion east and west, longitude was kind of essential. Now Huntsman happened to live near a glass works, and he saw the glassmakers putting in chips of old broken bottles, doing high-temperature remelts, and coming out with really great glass. So he tried the trick with steel. It worked, and there was what he wanted, the world's greatest spring. The point was, Huntsman's steel would also cut anything you could think of, so what it did for the lathe, and machine tools in general, and micrometers, and precision engineering, and steam engine cylinders, and the whole Industrial Revolution was something nobody could ever have dreamed of—least of all Huntsman, who sat there saying, "What happened?"

Sometimes the catalyst for major change will simply come in, totally unexpectedly, from outside your paradigm. Take the case of the compass. It came in from China via the Arabs in the twelfth century. Nothing much happened until Sir Francis Drake came back from over here complaining about the way the needle did funny things when you got across to this side of the Atlantic. Queen Elizabeth's doctor took time off (18 years) to look at why, and decided that the Earth was a gigantic magnet with poles. OK, so what? Well, to carry out his experiments, he built himself a lot of balls of various substances—lodestone, amber, sulfur, glass, and so on—to represent the Earth, so he could see what they did to his compass. As he busily rubbed these balls to make them attractive to his needle, he noted somewhat disinterestedly that sulfur was very attractive, and added a minor footnote to that effect. Around 1640 the mayor of Magdeburg in Germany, one Otto Guericke, read the aforesaid footnote and tried the trick again. While he was rubbing his sulfur ball one day to make it attractive, it cracked and gave off a spark, and—yes, you guessed it—electricity. From the compass. From China. Even if you'd spoken Chinese you wouldn't have seen that one coming!

One of the most common ways change is generated is through interaction between one factor and another, and usually in unexpected concatenations. Take the skills a goldsmith has. He's good at working soft metals and using molten alloys, and the hallmark of a good goldsmith is just that, his hallmark, the punch that puts his impress on his work. If you are capable of seeing that punched image in reverse, you can see how to cast a shape in the pattern made by the punch. And the pattern could be a letter, in metal, which is why printing was invented by a goldsmith—that's what Gutenberg was.

This interaction that can lead to change is often caused by imbalance, a kind of domino effect. The well-known modern one is that of the superplants. They give great yield—better than the old, less productive types. But they replace variety with a monoculture, and, if disease hits that, you've got no fallback.

That kind of domino effect—the knock-on effect of imbalance in one area upon another—gave us one of the major scientific discoveries in history. When cannons started being popular in the mid-fourteenth century, they pushed up the demand for metal, and that got people deeper into the ground than before. One of the things they found was that the deeper you go, the wetter it can get, and the old suction pumps wouldn't lift the water up higher than about 30 feet. Well, this problem caused all sorts of grief until one of Galileo's boys, called Torricelli, worked out that it had something to do with atmospheric pressure. A friend of a friend of his went up a mountain with a tube of mercury to see if pressures were different up there and down here. Well, they were, but what was the gap at the top of the tube full of mercury? It was the thing everybody said didn't exist—the vacuum. And suddenly you had barometers, airpumps, a new view of interstellar space, and a very different basis for science.

Now, the mechanism by which change can be generated isn't by any means always a technology-technology interaction. Take the ultimate effect of the telescope. When Galileo looked through it he saw satellites circling Jupiter. That blew a hole in the Earth being the center of everything, took humanity off its special philosophical pedestal, and prepared the way for a universe that wasn't arranged the way Aristotle had said, but the way Newton was to say—like a giant clock, running by itself, with God maybe long gone on other business. Religion took a knock from that from which it never fully recovered.

The German mathematician and businessman Gottfried Leibniz, working on the planetary dynamic problem at the same time as Newton and looking at the kinds of mathematics you'd need to measure infinitesimal rates of change in movement, decided that he had his hands on a tool of cosmic philosophical significance. If you could measure *that* infinitesimally, were you getting to be able to measure the basic units of existence? If you were, said the philosopher Immanuel Kant a bit later, you could discover and measure the way all things shaded into all other things at that scale. The new philosophy became known as *naturphilosophie*.

Its concept of "oneness in all" spawned romantic poetry and music, nationalism, and revolutions including yours. It won't surprise you to know that Jefferson was a *naturphilosophe*. *Naturphilosophie* also helped to bring about modern medicine. In 1810 a French surgeon named Xavier Bichât, another follower of the new philosophy, went looking for the vital, infinitesimally small bits in his business, and found body tissue—20 types of it. Incidentally, he set the fashion for grave robbing and also noticed that, if you were sick, changes showed up later on in the tissues of your unfortunate corpse. Maybe these happenings could be correlated. So pathological anatomy was born, and with it the modern idea of disease as a localized phenomenon, as well as the opportunity to look for, and find, bacteria.

Canal building, spurred by the transportation needs of the industrial revolution in Britain at the beginning of the last century, turned up fossils in the strata they were cutting through. However, mysteriously, some of the fossils were of animals that didn't exist any more, and most of the strata didn't have fossilized humans in them. Well, here was a big problem! God was supposed to have made everything at once, during creation, and yet here were some animals that obviously had failed, and no people back at the beginning of everything.

So what was this—mistakes by God, with some things having been created later than others? Well, you know where that led. By the time the geologists had finished discovering that the extent of the entire fossil record through time was like an eyeblink compared with the age of the Earth, it was a simple matter for the whole thing to be organized into a new view of things by Darwin. This brings us to the materialist, physicalist world we live in today, where people maybe aren't something special created by God in his own image but just a pile of chemicals.

This is also a world where one interpretation of what Darwin meant by "the survival of the fittest" boosts rugged individualism and makes life not so easy on the underprivileged, and where his idea of "perfectable humans behaving according to laws like the rest of the organisms in nature" and, therefore, being part of societies that perhaps can be "changed for the better" is alive and well and regimented in the Siberian labor camps.

I went through all those examples of change in action at length to give you a good idea why, when paradigms start to shift, the unexpected way they go is a shock to the system. This is why any time you do manage to produce a way of thinking or of doing things that seems to work well, you hang onto it. If you can work out a way to maximize what your society can do with the tools at its disposal—give it the widest flexibility in terms of individual action and at the same time protect it from random, maverick action with some kind of rules—great. That's why the institutions survive; they're set up with the tools of the time and they're systems that permit routinization of the group's operating problems so the individual members can get on with working or having fun while the institutions handle the day-to-day running of the place.

So we keep the institutions that appear to do a good enough job because it's easier than handling the problem of assessing how well whatever new tools you might have come up with could handle the same basic daily problems in radically restructured institutions. So never mind if the institutions don't quite fit the new paradigm you're moving into with your new tools; it's better than experimenting. Corn, after all, is only corn because it's stood the test of time. So most of the institutions we live with are, in some aspect or another, anachronistic.

Take the law. Cross examination originated far from the courtroom, as a teaching technique in eleventh-century Italy for making sense out of old manuscripts. The technique was called glossing. Another institution, the language I'm speaking now, effectively froze when it was printed in grammars in the fifteenth century. The modern university started life as a place in twelfth-century Bologna designed to train lawyers to handle jurisdictional and property cases, particularly between the Pope and the Emperor. Many of the internal structures of our universities, at least in Britain, remain unchanged since that date. Representational government is something that was thought up in the eighteenth century when only the foolhardy few would risk the mud and the bandits to get to London or Philadelphia.

We hang onto institutions as if they still meant what they did originally, as if the paradigm in which they originated hadn't shifted. We accept politicians talking about what they can do to the economy as if the world still consisted of independent, separate sovereign states whose acts had no effect on each other, or as if the meaning of fundamental beliefs had not changed. One good example is yours: "Life, liberty, and the pursuit of happiness" means nothing like what John Locke meant when he thought about it in the seventeenth century. Our freewheeling adaptation of it today would have shocked him rigid. "Liberty" for Locke meant knowing and accepting where you stood in society and sticking to the rules that governed social class mobility, such as it was. "Happiness" meant amassing property and riches without being bothered by government! He would have thought we were living in anarchy.

Be that as it may, we regard the institutions and their associated slogans as helping to preserve cohesion and stability in our paradigm—except that this is a cohesion and stability which is, as you've seen, at best transient. Once the French philosopher and permanent exile René Descartes got his hands on the way we in the West thought, stability and permanence went out the window. Before Descartes and his seventeenth-century paradigm shift, you said *credo ut intelligam*—I believe, and through my belief I come to understand. After him you switched it around: *intelligo ut credam*—give me the facts and I'll let you know. In his great *Discours Sur la Methode* (or, "how to think"), he gave us the modern approach. He called it methodical doubt. He said, "If they tell you it's certain, call it probable. If it's probable, call it possible, and if the deal is that it's possible, forget it." And Cartesian methodical doubt is the engine of the modern scientific world and the bringer of accelerating rates of change.

So, where have we ended up? If the mechanisms of change are as serendipitous and as hard to second guess as I have suggested—and we are, thanks to Descartes and others, in a world of increasing rates of change—are there any lessons to be learned from the past to help us at least adapt? Is it true that those who are not prepared to learn from the past are condemned to repeat it? Do we really only know where we're going if we know where we've been? Well, there are repeating factors, back then, that seem to be present when change occurs, much the way cholesterol is with heart attacks, present and only maybe causative. First there's the one that appears to be the most obvious, that change happens because you need it—"Necessity is the mother of invention" and all that. There's an interesting study of Europe up through the late Middle Ages that seems to show that innovation happens and is taken up most in areas of marginal circumstances and stress, and least where things are pretty comfortable.

Let's look at the ancient Egyptians. When you've established the simple fact that once a year the Dog Star, Sirius, appears just before dawn (after having been invisible for seventy days) and one day later the Nile floods and dumps fertilizer and water on the land, and that it does so with extraordinary exactness every year, you develop a calendar just to tell you which day Sirius is going to appear, dig your irrigation canals, and sit back. That's all you need in the way of new tricks, so Egyptian society never changed after that initial step. It never needed to, in 3000 years.

But the ancient Greeks? Well, put yourself in their position. In the eighth century B.C. you live on narrow coastal strips in what is now modern Turkey, in little city states with just enough to survive on. The weather is lousy and uncertain, and the barbarians are clobbering you with regularity. You've got to get out and trade, make a buck, just to keep going, so you think up ways of systematizing the method of hustling business. You look up at the sky, and what you see is not Sirius rising and nothing else; you see a great road map for your seaborne traders to use. You work out star tables to navigate by, and the more you look, the more you see that the permanent perfection of the night sky is a lot different from the temporary mess down here. So curiosity becomes a way of life. No wonder the Greeks invented their particular form of curiosity. (They called it *philosophia*.) It's what you get when you're looking for answers.

In a sense, it was Greek philosophy, born of their difficult circumstances, their desire for answers to questions, that started change happening in Western culture. What got it accelerating, though, was something else, and that's the ease with which people communicated, moved ideas around. The easier you cross-talk, the faster change happens.

Take medieval Europe. When the Vikings and the Saracens and the Hungarians stopped the rape-and-pillage stuff in the tenth century, people started coming out of the woodwork and building little roads toward each other and traveling along them. The next thing you know, you got the medieval water-powered industrial revolution, which kicked the European economy into high gear within three generations.

In the Renaissance, a hundred years after the arrival of printing, you had 20 million books, most of them in specialties that could only exist when the specialists had a way of reading each others' stuff. This gave us nothing less than the scientific revolution of the sixteenth and seventeenth centuries, and a slew of people talking the kind of incomprehensible stuff most of you live with in your area of expertise. Don't be insulted; how much do you understand of the language of paleontology? Specialization is essential. I'd hate to have flown here in a plane designed by a plumber.

To return to my point about communication, in the nineteenth century, after the development of electromagnetic systems for moving messages around (i.e., the telegraph and the telephone), the whole body of modern science emerges—in particular, physics.

So, the ability to communicate seems to be a basic factor in the mechanism of change, and we have communications today that make earlier forms look like hieroglyphs painfully chiselled out in stone. New developments in areas like magnetic bubble domain memories and superconducting materials will enhance our ability to use data beyond anything we've even begun to think of. With our present facility for communication, we're doing more of one particular trick than at any time before. And that trick, it seems to me, is putting things together.

Let me suggest a new axiom: juxtaposition is the spice of life. Humanity's biggest talent, unique to us, is juxtaposing, finding and operating novel relationships between things or ideas. Indeed, at the turn of this century in Vienna, a group of thinkers who were to have a profound influence on Einstein (the positivists, led by Ernst Mach) came to the conclusion that all science could talk about was relationships. This was after Michelson and Morley had failed to find the ether. You remember that up in Cleveland in the 1880's the two of them were looking for a medium that would be the carrier of light, magnetism, and electricity. Everybody called this medium "ether." Well, Michelson and his friend were trying to show that the two halves of a split light beam would come back together again, out of phase, because one-half had been shot in the direction of the Earth's travel (against the so-called stream of ether) and would take a while, and the other, which had gone perpendicular to the ether, and so wouldn't suffer drag, would return early. Actually, there had been no difference at all. The beams arrived back simultaneously.

Then Fitzgerald, in Dublin, made things worse by saying that this was because the forward motion of the Earth was contracting one part of the instrument exactly the right amount to give the

17

forward-moving beam a shorter route to travel, so it could get back exactly that much faster, and match its other half. And, the experiment could never be carried out without that happening. So, whether or not there was an ether, you'd never be able to find out, since all you could get would be the local-effect results from your work. That led the positivists to state that science could only ever produce relative, not absolute, results. All you could talk about was relationships.

But I mean relationships in a rather more limited sense—the sense of the way properly original thinking involves juxtapositions that have never happened before. Of all the mechanisms of change, it seems to me that this is the fundamental one, the on-the-spot local "fitting together" of disparate phenomena that comes up with the kind of changes I've been describing. This ability to juxtapose is not a very surprising one; it's logical enough in the light of our own neurophysiology.

Recent ideas on neural activity suggest that the brain operates in a very associative way, with small neuron clusters containing core concepts, rather in the way a battery holds a trickle charge. These core concepts would be irreducibly small fragments of sounds or sights, or any phenomena that you experience. And these clusters are all, in some way, apparently interconnected, set up in microcolumns and macrocolumns, each column made up of· millions of these little clusters of neurons. Now, if you consider that the brain passes information by means of synaptic junctions (the bits where one neuron almost touches another) and that there are potentially more of those kinds of connections in the brain than there are atoms in the known universe, you get a feel for the immensity of the network. With this associative system, to retrieve data, you go in, so to speak, anywhere on the network and find the target by association. Given the scale of things, an associative approach might be the only way the whole huge complex could work. Anyway, retrieval by association would be a good survival mechanism, because it would make you very flexible.

The other interesting thing about functioning in that associative way is that as you head along the associative links toward the target, you may become aware of other core clusters that you weren't aware of before, because in a sense you simply drive through them. That, in the simplest sense, would be why the brain is capable of associative chains of thought like this one: look, see, water, glass, mirror, image, painting, oil, Arabs, desert, sand, castle, and so on. It's why poetry works: "What oft was said but ne'er so well expressed."

Jokes appear to work like that too; the punch line makes an association you hadn't thought of before, and you laugh because you didn't get to the new associative link before the person telling the joke (which is bad for survival, but, as it turned out, you weren't in any danger). Let me try what I mean. Take the concepts "bird" and "fruit." All of you have those concepts associated in your own personal way in your network. I don't know whether it's like mine, with "bird" and "fruit" associated by "trees," but let me see if I can put those two concepts together in a way that they've not been put together in your brain before, and we'll see if my theory works.

A drunk goes up to his host at a party and says, with all that clarity used by the very small and young: "Excuse me. Do lemons whistle?" To which his host replies: "No, lemons don't whistle. Why do you ask?" And the drunk says, very chagrined: "Oh. In that case, I have just squeezed your canary into my gin and tonic."

You see what I mean. What I'm saying is that the basic mechanism of change—the juxtaposition, in a novel relationship, of apparently unrelated phenomena— may operate in the same way a good joke does! It may also be why change is almost always so serendipitous and unexpected—and hard to forecast.

Given all I've said so far, let me be extremely speculative. We've seen how the model I'm talking about functions, and some of the ways in which it's changed. We've seen how difficult it is from within the paradigm to see why moving to a new model would be beneficial—"Better to keep the devil you know."

We've seen that when paradigms are about to crack, there's generally some social unrest going on. In Copernicus's time it was barbecuing freethinkers (they called them heretics). With Darwin it was supposed to be the end of beliefs and standards. In the thirteenth century they said paper would devalue the words written on it, and for Gutenberg it was, "Printing will take away our memories."

With us, it's all the words you see in the media: alienation, frustration, me-generation, immorality, illiteracy, and so on. So is our paradigm about to go through some of the agonies I've been describing? Is it due to shift? Well, obviously it is. But let me suggest that instead of moving to a radically new paradigm, we may, because of the tremendous facility for interaction that communications gives us, be moving to a no-paradigm culture.

If a paradigm is, and has always been, a structure built on an agreed core of common beliefs, knowledge, value judgments, social constraints, and so on, then are we heading the opposite way, to a situation of no common agreed center, of shifting, pragmatic local standards, with failure of what we used to call consensus and regionalism globally on the increase again after the early years of *Pax Americana*, with the nation-state obsolete, and so on? We would be a society physically and psychologically fragmented, because with soft energy options and telecommunication, "centralization" and "economies of scale" (those catchwords of the last years of the Industrial-Revolution paradigm we're coming to the end of) are no longer necessary.

To those of us condemned to repeat the lessons of history because we won't learn from them, what I'm describing sounds like a frightening prospect. Chaos is what it sounds like, but isn't what's happening just a paradigm shifting (like all the others did) because we're ready for the shift? Change occurs ultimately because we want it to. We have the tools because at some time we decided we wanted them. These new tools, provided by science and technology, are more than just tools—they're instruments of social revolution, violent or peaceful. As the tools change, so too does the ability of society to organize itself.

Once we needed god-kings, or feudal lords, or absolute monarchs, or no sex before marriage, or empires, or 12-hour days, or whatever, to keep ourselves together. Now, maybe, we don't need centralized social structures and rigid regulatory mechanisms any more. We are, after all, as Immanuel Kant said, creatures of the imperative. If the ethics start to get in the way, we dump them.

But let's take a brief look at the kind of behavioral social dumping we may be up against with some of the possible results of our newfound abilities to initiate change much more readily and rapidly because we can juxtapose things inside the computer, where we have a facility for juggling the mix like never before, at a rate and in volume almost astronomical. And, by the way, for those of you who feel nice and safe because of the old sayings "Garbage in, garbage out" and "A machine is only as good as the people punching the buttons," try some of the newer heuristic systems that learn from their own experience.

The main thing, it seems to me, is to remember that technology manufactures not gadgets, but social change. Once the first tool was picked up and used, that was the end of cyclical anything. The tool made a new world, the next one changed that world, the one after that changed it again, and so on. Each time the change was permanent. Using the tool changes the user permanently, whether we like it or not. Once when I was in Moscow talking to academician Petrov, I said, "Why don't you buy American computers to get you into space quicker and more effectively?" He replied, "No fear; they'd make us think like Americans."

You only have to go back a few years in this century to see how our gestalt, our way of behaving, our values, have been changed by science. If I say just a few names, you'll get my point: the Pill, calculators, jet airplanes, television. Take those examples and look at their secondary social effects. Yes, the Pill has made family planning feasible, but now the Third World regards it as a suspicious imperialist Western trick to keep their numbers down while we go on with our "economic imperialism."

Calculators have changed the meaning of testing people in certain kinds of knowledge, which we need to do to ensure publicly accepted standards of professional ability. Jets mean people can now fly and visit the ends of the Earth, but they also mean that we export our way of life and our sometimes unacceptable value systems to places that neither want nor need them. Television makes my life one of totally vicarious experiences. It gives me packaged glimpses of the world beyond my horizons, takes away my comfortable preconceptions, and replaces them with glossy, quick-fix substitutes that are even less good to me than my preconceptions were. All I know now is that I don't know!

To get back to my "dumping" idea, you see how the gadget changes more than just what the ad says it will do. With our rates of change, the only constant in our paradigm may well become change itself. All you can be sure of about tomorrow is that it will go on being different, and, if you're lucky, only at the same accelerating rate.

Above all, the judgmental systems from the old paradigms may not work in that world. Today we are, in fact, the last of the old world, living with institutions that are already creaking, facing twenty-first-century problems with nineteenth-century attitudes. Most of us find difficult to accept what we might have to dump. We face questions like these:

If criminality is caused by XYY chromosomes, who do you blame for a crime, and why do you punish at all?

When everybody has a home computer work station, what happens to unions, the infrastructure that runs the roads and transportation systems, the community life that "work in a central location" means, the new isolation of being alone most of the time?

If data banks carry all the knowledge we possess, to be accessed at need, what will be the purpose of memory, of "knowing" anything? And what happens if what you got from the machine yesterday (what we'll call "what you know") is different when you go back to the machine today?

If you have no expertise because expertise is no longer necessary, what are you left with?

If technology provides virtually free energy, with the ability to turn anything into anything else (which we can already do—it's just too expensive to be feasible), and we no longer need the raw materials we used to because we can now make them, what happens to the materials producers in the Third World? Unlimited energy, the so-called philosopher's stone, brings far more questions than answers. Not the least of these is the new importance it will have for the planetary heat budget, which at the moment is pretty much only the business of nature.

Well, my guess is (and here I remind you of the unquestionable value of any guess made from within the inevitable limitations of our paradigm) that we're all headed for one of two kinds of future.

In one future, we take on the new data systems the way we took on all the other tools in the past, with a view to making them do what we've always done up to now, only better, faster, and cheaper. In this case, I think we're in for a dose of Luddite reaction as our social structures fail to take the strain of that much shift that fast in the working habits of the population, not to mention the redundancies that come if all you do is replace people with machines. The other problem with that old-paradigm approach is, of course, that you do what Bell up at Yale says, and turn into a two-class society. You have the numerate, who have access to and ability to maximize use of the data systems, and you have the leisured serfs, who don't, and who get paid for a 10-hour week with nothing to do but wish they knew how to use their spare time.

It has often been said that the public doesn't appreciate the speed with which things have developed in data systems. I like the analogy that if your Rolls Royce had done what computers have done over the last 20 years, it would cost a dollar and do a million miles to the gallon. People, I think, just don't understand the velocity with which this new post-Gutenberg era is coming toward us.

The other future I mentioned is a good deal more difficult to forecast. It's very much up in the air. All I can do is to be extremely speculative again. I suppose what I'm suggesting is a crash restructuring of the educational system. I've been a teacher myself, so I know how easy this is to say and how difficult to do. However, if we were to manage some kind of interdisciplinary curriculum that taught people not the facts, which would be obsolete before they used them, but how to use the data systems to juxtapose, to look for relationships in knowledge, to see patterns in the way things happen and affect their lives, then perhaps we would be moving toward a very different type of society, one free of a central paradigm at all.

After all, the only need there ever was for a paradigm was based on the strictures placed on society by its contemporary tools—or rather, lack of them. Now we have a tool—electronic data systems—that could lift almost all of those strictures from us, that could create a society that might be pluralistic in the extreme, lacking in any of the virtues we now ascribe to concensus, materialistic in every sense, highly articulate, what we would call unethical and immoral (what it would call pragmatic), self-sufficient (what we would call isolationist), libertarian (what we would call permissive), and above all, open-minded, curious, and tolerant.

Sounds like a weird mix? Well, you asked me here to speculate. But in one sense it's what we've been heading for all along—a kind of controlled anarchy, kept in balance by the electronics. It's the truest version yet of what John Locke meant by "the unfettered pursuit of happiness by every man." And if the vision bothers you, remember that once we decide that the paradigm is shifting, we adapt extremely quickly. Your great-grandmother, after all, would have thought you a drug addict for taking an aspirin.

Question: You discussed the future paradigm as perhaps being nonexistent. Is it possible that the paradigm might be evolving just as human evolution evolved to the point where it is reaching its own sense of oneness with its future? In other words, we are part of the paradigm and the paradigm is what is evolving. We are part of the evolution.

Answer: The great thing about that question is that it's unanswerable. I mean, by definition it's shear speculation again. All we can do is talk about it because we're inside the paradigm. These wild speculative guesses are set in concrete because they're within my paradigm. If they sound wild you should hear what happens if you come from another planet.

Question: At the end of your talk you quoted Locke and said we should all seek our own happiness. Happiness is a paradigm. We all live in dreams. Every person has his own idea of what happiness is. We have paradigms that are imposed on us by the world, but we each have our own paradigm. I don't know what human life means without dreams that are paradigms.

Answer: Yes, I think that's very well said. All I was suggesting was that this might become more possible than in the past. I didn't mean that you lose your paradigm. I meant that perhaps your paradigm becomes a little less constrained by everybody else's paradigm.

Question: As scientists working for the government we are often asked to forecast what new inventions we might come up with over the next year. I wonder what implications that has for our role in bringing innovation into the world.

Answer: I think it's a superb example of what I was talking about. The government decides to make you decide what you're going to discover, and if you don't come up with it you lose the grant!

Question: Of all the countries you've worked in, which one, in your opinion, provides the best education, and, in particular, how do you view education in the United States?

Answer: That sure sounds like a quick way to get my head chopped off! I think educational systems tend to be structured according to the societies in which they work. I mean, our educational system in England is extremely difficult, different from yours, and very elitest. A very small percentage of us go to university, and we're used to choosing the subject that we study at university at the age of 16. We specialize in only two subjects from ages 16 to 18, and we then take a national examination in those two or three subjects. Only one of those subjects is what we go to university for, if we pass a competitive examination to get a place at the university, and the ratio is usually about three or four hundred people to each place. Now, we have to have an elitest educational system like that because we are very small and we've become quite poor ever since we lost the jolly old empire. If we didn't have that kind of high-quality turnout we wouldn't have enough people producing enough stuff on the market for us to sell anything to anybody. So I think we have an elitest educational system not because it's a hangover from the old imperial days, but because if we don't produce a very, as it were, sharp-edged elite intellectually, we won't be able to compete with giants like you on the market.

Question: I would like to ask whether the, what shall I say, elite in Britain and perhaps in Western Europe believe in full employment not merely because of the necessity for having the things that people produce when they're fully employed, but rather as occupational therapy for the masses, around the idea that idle hands do the devil's work, and that whereas intellectuals can keep their minds occupied and out of mischief, the common man is not capable of this. George Orwell said something like this (and it's not something I agree with), but I would remark that Eric Hoffer said the common man was lumpy with talents and could do all kinds of things besides produce goods and shouldn't be viewed merely as a production machine. Can you speak to that, sir?

25

Answer: Well, I can't speak for all of Europe, but I think the French probably think that full employment's essential and they've had four devaluations of the franc as a result. It seems to me that full employment is a relatively new phenomenon. We've slid over into economics, and I'm extremely worried—I think anybody with any sense and honesty always is, in that subject. However, I believe I'm right in saying that full employment is a twentieth-century phenomenon. The concept didn't exist to any great extent at all prior to that. And I think it probably came at the tail end of a very healthy, burgeoning post-Industrial Revolution in both America and Europe. I think what we're seeing now is a transition period to what Bell calls a post-industrial society, and it's a period aided and abetted, of course, by the recession, which is caused not by the fact that we can't switch paradigms but because oil costs a great deal. I think the situation, fortunately for me, is so confused that no clear statement can be made on it by me or anybody else except a politician.

Question: If I understand you correctly, it seems to me that you're putting out the impression that our technology is running away from our society. In other words, it's speeding up at a rate that we can't quite keep up with. In the past, when this has happened to societies, some major upheaval has occurred, whether it be sociological or financial, economical, or revolutionary, like wars. Do you have any idea what is going to cause us to catch up with our rapidly advancing technology?

Answer: Well, I think part of what I said earlier indicates what I think about that. First let me just dispel any idea that I believe in the so-called force of technology. I mean, technology is what people do. You invent the tool because you want it, or because you perceive an imbalance or a need, or you're just greedy. You say, "I want this piece of technology," and it comes into existence and you use it. I think society gets technology as it gets governments that it deserves. Sometimes, but not very often, technology tends to go a little faster than our ability to keep up with it. I'm not sure that this has happened to any great extent in the past, but I'm sure that it's about to happen now. I think anybody with any sense would recognize that electronic data systems are going to make a quantum leap in terms of the effect of juxtapositioning, as I said earlier. As to what we can do about it, it seems to me that the only way to get into it is through the educational process. It's too quick, and you can't have a quick-fix answer. It's no good teaching us what to do. I think you've got to begin with the children who are 4 years old now and start the process there. As I said, I just hope some teachers who are better than I am at organizing this kind of thing in education, which is tremendously difficult, will get on with it, but I can't see any other way of doing that. We are up against a period of very difficult transition.

Question: Being somewhat of a video game fanatic, I've noticed that extremely small children play video games much better than anyone else. They're well adapted to the electronic age because they have far fewer preconceptions, apparently. The way things are going, it looks as if things are going to get less and less expensive and more and more reachable in terms of the spread of technology and the spread of knowledge. Everyone can learn. Even if we can't feed everybody in India, we can teach them all how to read. Pretty soon everybody will have his own terminal. Now, over the years, one of the major complaints of the Third World, even the Third World in the United States, has been that they never had the chance to get a leg up because they were deprived from the start. So, could it be possible now that we really will achieve a parity of sorts because everybody will have the same chance once this technology becomes more equally spread?

Answer: Well, it depends entirely on what regulations are applied to the use of the technology. If I live in a totalitarian state and I produce a computer you can bet the people who use it are going to use it in a very different way than they use it in Spokane. As Petrov said to me, "I don't want American computers; they make us think like Americans." So, first of all, I think the thing you're discussing is a matter primarily of social and political consequence. Technology doesn't do things to us; we, I hope, do things to technology. So, the fact that every man has his own computer does not necessarily mean that we will have instant parity and literacy everywhere and a bright and happy future, because there are a lot of governments that would like to make sure that half their population only ever plays video games. Keeps them quiet. It's no good having all the terminals in the world unless you've been taught to ask the questions. If you're given all the knowledge that ever existed, where do you start? How do you know how to structure your approach to the knowledge? I don't think that the mere existence of the games presupposes that we're on the verge of that kind of understanding.

Question: Mr. Burke, in your book *Connections* you speak about the development of fusion power, which could provide unlimited energy, yet it also gave us the bomb. You seem very optimistic about the future as it applies to free peoples, let's say Western Europe and America. What do you see in the future for change in the Soviet bloc, or in Russia?

Answer: I don't think I'm blindly optimistic. I think we have a great survival capacity, we human beings; otherwise there wouldn't be a crowd of us in this room tonight. We've survived an awful lot of vicissitudes. So, in that sense I'm a long-term optimist, because I believe in evolution. In the short term I'm not too sure whether somebody's going to put a nuclear device downtown somewhere in the world and that will be goodbye everything, possibly. So I'm not sure that I'm just a straight optimist in that sense. I think the business about what would happen in societies that are less "free" relates to what I was saying earlier about the kinds of societies that Professor Bell suggests we might be coming up against, in which the technology is a marvelous way of imprisoning people. You can as easily imprison people with it as free them. Go back to Petrov. They're going to make their computers to function the way they want them to because they don't want to have to start thinking like we do. In this sense, your question does tie in to the previous question in that the technology by itself is nothing; therefore, what you're talking about is again a political thing. You can as easily imprison people with computers as liberate them. It seems to me that's all one can say about it.

Question: With your rich background in the development of technology, I'd like to know if one thing that I think that I've noticed in my lifetime has a precedent in history. When I was a kid I remember that cars had a top speed on the highway of 55 mph, and then they souped up the highways and the cars and started going 65 mph, then 70 mph, and then there was a choice to go back to 55 mph because it wasn't safe and because it used up too much fuel. On the issue of nuclear power, even if there hasn't been any sort of decision as far as moving away from nuclear power and the other sorts of technologies that have come out of it, at least the dangers have become very much more a question now than they ever were. It's not that a technology came in which was questioned and then accepted; it's a matter of a technology that came in and was accepted first and then questioned. I'm just wondering if that's something new.

Answer: Technology assessment, as far as I can see, is quite new, yes. I think it has to do with the fact that we are more capable of discussing these matters among ourselves than we ever were before. I think that when two or three people are gathered together they start to ask questions. I don't think that has been the case before on this kind of scale. I can think of inventions that came ahead of their time and weren't used, but not as a result of cogent argued discussion among members of the community, because there really wasn't any such thing before 1900. My favorite invention that didn't get anywhere was Voltaire's electric gas bomb. It had a spark igniting this *mal aria*, what they called bad air, but nobody really did anything with it until the spark plug came along, under totally different circumstances.

Question: Could you speculate as to whether or not there could ever be communication, and I ask this technically, between the hearts of men? I'm very nervous about how you'll take that. In view of the confrontation we have between East and West and the nuclear aspects, all the communication we have today doesn't seem to be breaking through some problem area there. This is very philosophical or theological, but if you care to speculate on it I would appreciate it, sir.

Answer: I don't know how you encourage people to have one heart talk to another. I presume you don't mean the pump, you mean more than that. I'm always attracted to the fact that the only purely democratic, purely objective, purely self-regulating, purely honest activity known to mankind is science. Everything else is lies and partial views. There is nothing you can do about the fact that when I drop this piece of paper, it falls. You can't tell lies about that. I can't think of how you can tell lies about it. And it seems to me, therefore, that I've gone into this spurious, superficial television approach to try to tell people about science principally because I believe that science is the truest possible route to what you do want to happen.

29

I remember once we were having a discussion in the BBC when there was regionalism in Britain, when Scotland was going to secede, and the BBC began seriously to consider maybe moving some of its offices out into the provinces to find out what the grass roots thought. Well, somebody killed the idea by saying, "If you tell me what the difference between Edinburgh physics and London physics is, I'll move to Edinburgh." It seems to me, as I say this, that the way I describe science gives me the hope that it's the only way one can survive because everything else has to be persuasion, partial view, opinion, and belief. But nobody can alter the fact that this page drops when I let it go.

Accomplishments of Science by the Year 2000

Jules Bergman

Jules Bergman

As the first full-time television network science editor in the US, ABC News Science Editor Jules Bergman covered all 37 manned flights in the US space program, beginning with the original seven Mercury astronauts and the Space Task Group at NASA Langley Research Center. He frequently participated in the astronauts' rigorous training programs and flight simulations. He also covered the historic US-USSR joint Apollo-Soyuz Test Project and the fall of the American *Skylab*. In the 1960's, when the US was deeply involved in its emerging space program, Bergman's reports were the main source of information on this subject for many Americans. A pilot himself, Bergman has also covered the first flights of almost every new US military and commercial aircraft as well as major airline disasters around the world.

In the 1970's, Bergman turned more of his attention to documentary work, winning an Emmy Award for his contributions as co-writer and narrator of the ABC documentary *Closeup on Fire*. He has written and narrated documentary programs dealing with the energy crisis, sports injuries, aircraft and automobile safety, the hazards of asbestos, and nuclear power. In the spring of 1979 Bergman contributed to ABC's coverage of the nuclear power plant accident at Three Mile Island.

In the field of medicine, Bergman covered the beginning of the transplant era as well as various new developments in cancer treatment. More recently he reported on the swine flu controversy and the "legionnaire's disease"mystery for both the ABC Evening News and ABC's *Good Morning America*. He occasionally hosts ABC's Sunday afternoon interview program *Issues and Answers*, questioning guests from the fields of science and space. Bergman has won many awards for scientific journalism and has written numerous articles on space and science for such magazines as *Readers' Digest, The New York Times Magazine*, and *Esquire*. He is a native of New York City and attended City College of New York, Indiana University, and Columbia College. He completed postgraduate studies at Columbia University, where he held a Sloan-Rockefeller Advanced Science Writing Fellowship.

Accomplishments of Science by the Year 2000

By the year 2000 we'll be flying on supersonic transports or, more likely, hypersonic transports. We'll have put the genetic code to work and begun to engineer out congenital birth defects and inherited diseases. We'll have made a real start toward the prevention and conquest of both heart disease and cancer. The normal life span should be 85 to 90 years, and anybody suggesting mandatory retirement at age 65 will face unthinkable punishment. However, these advances require that this country correct its inadequacy in science and technology. We are living off the past and ignoring the future. Other industrial countries now turn out more engineers and scientists than we do. The United States needs a manned space station—it may be the most worthwhile investment in future technology and science that we can make. We should also start planning for a manned Mars mission. Of course, it would be so expensive that no single nation could afford it, which is probably good. It would then become an International Mars Mission and would involve at least the United States, Western Europe, Japan, and very possibly Russia. Such a joint mission could pave the way for world disarmament.

As a society, we walk a tightrope between limbo and extinction. We're on a threshold of survival, in a society threatened as never before to find the way, with less and less margin for error. The decades ahead to the year 2000 and beyond, as were the decades just past, can be either interrogative, presumptuous, or insane. And we have to create our own flight plan, because this Earth didn't come with one telling us how to get to the future safely. The winds of change are blowing across this land. We're a nation that can't afford many more crash landings, yet we keep putting untrained pilots in the cockpit and thinking we'll all somehow come out all right. No way. No longer.

33

Today's climate of controversy, confusion, and too often outright hysteria, compounded by exaggeration and often lies, has brought us to a real crisis, a crisis of confidence in ourselves, a crisis of credibility, and we're all close to totally disbelieving one another in this country, the media and the government, the people and the government, and sometimes, although not enough, the people and the media. You don't ask us enough questions. The day's news leaves even us in the news business feeling stupified, outraged, and helpless. The whole thing reminds me of the 91-year-old astronaut who was asked for his views on homosexuality and replied, "They used to hang men for it in my grandfather's day. In my father's day, they put people in prison for it. Now it's permitted. Well, I want to get out of here before it becomes compulsory!" Or, as one researcher friend puts it, "It's always darkest just before it gets totally black." The lesson of the past, if there is one, is that we can't ignore the future. It is here, now, and inescapable.

Right after Sputnik, 25 years ago, you may recall we were warned that we lacked enough engineers and scientists, so engineering schools were enlarged. Classes doubled, and the liberal arts and humanities were sacrificed when they needn't have been. Now we're entering a new era, the era of microminiaturization. After a lengthy study, the National Academy of Sciences (NAS) warned that "we're raising a new generation of Americans who are scientifically and technologically illiterate." Other countries, specifically Russia, China, and Japan, put much greater emphasis on math and science education than we do. Russia graduates 300 000 engineers a year; Japan, with half of our population, 75 000; and the US, 60 000. The number of class hours spent on math, science, and engineering in those countries is about three times greater, according to the NAS report, than those spent by even the most science-oriented students in the US. That report has already gone to each of the nation's 16 000 school districts and school boards. As for the Japanese, they're so far ahead of us in many areas that they actually have the answers to our problems before we know we have a problem.

It's not all bad news technologically; the lesson of the past is that we underestimate the future. The history of science is loaded with underestimates, short-sightedness, and skepticism. In the March 1904 issue of *The Popular Science Monthly*, a famous engineer reviewed the first flight of the Wright Brothers at Kitty Hawk. Half the newspapers neglected the event, thinking the Wrights pure nuthatches and the story too ludicrous even to deserve mention. If there had been television, we would have covered it just to show you the crash. The famous engineer, Octave Chanute, looked deep into his crystal ball and predicted "that such flying machines may even carry mail in special cases, but the useful loads carried will be very small. The machines will eventually be fast, they will be used in sport, but they are not to be thought of as commercial carriers."

A new climate of almost hysterical disregard for the technological needs of an overpopulated world has sprung up, a world which would starve and succumb to disease if we tried to return to the simple life of a century ago. Technology must not be destroyed; it can and must be controlled, and not with distortions leading to erroneous conclusions, not with unproven charges. As Admiral Rickover once put it, "Half truths are like half a brick; they can be thrown farther."

Let me give you two examples of leaping to conclusions without the full facts. Back in the 1890's, a certain California newspaper was apprehensive about the harmful effects the railroads would have on the environment. If the trains crossed the Mojave to get to the Pacific, this newspaper editorialized, "the huge iron rails will reverse the Earth's magnetic field with catastrophic effects." Now that's real science! One hundred forty years ago, the Royal Society in England warned against the railroads, claiming that at speeds over 30 mph, the air supply to the passenger compartment would be cut off and people would die of asphyxiation. And the college of physicians in Munich, for its part, warned that at 30 mph, travelers would suffer headaches, vertigo, and possibly lose their sight because of a blurring effect. Over 30 mph, great catastrophes were predicted, because everyone knew that even a twig would shatter the wheels. In 1936, to come closer to the present, a War Department colonel visited Robert Goddard, the father of American rocketry, who had by then clearly demonstrated the practicality of the rocket. The colonel, doubtless fresh from the calvary in George Custer's historic triumph at Little Big Horn, dismissed Goddard's work as "sheer poppycock of no practical use in modern warfare." After World War II, Wernher von Braun said that it was Goddard's work that led the Germans directly to both the V-1 and V-2 rockets. It was Goddard, by the way, who left us with this legacy: "It is difficult to say what is impossible, for the dream of yesterday is the hope of today and the reality of tomorrow." To put it another way, as Santayana did, "Those who do not learn from the past are doomed to repeat its mistakes."

We in the media have learned to view prose, projections, and predictions with a jaundiced eye. As a famed aircraft accident report concluded, and I cite this to my engineering friends all the time, "Extrapolation is the fertile parent of error." Enough of the reports and predictions that have crossed my desk in the last decade have suffered from exactly that fault to teach me to be supercautious. As Pogo once said, "We have met the enemy and he is us." To which we might add, we have faults we have hardly used yet.

The Space Shuttle is the beginning of an era, one of our practical utilization of space as well as of its exploration. I envision our use of space to be about one-third commercial, one-third military, and one-third scientific. There are, by the way, nearly 400 communications satellites already in orbit. They have, of course, revolutionized our society, giving us everything from cheaper phone calls over longer distances to better TV signals. When the President promised us that the era had come to safeguard our security, he was referring to Star War type lasers and particle beam weapons, which, as you know, are already in progress. Lasers are working experimentally and have shot down drone planes and missiles and bored holes through the solid steel sides of target ships. Particle beams may be just a scientific figment of fiction, but can we afford that risk when the Russians are fast moving ahead in their development? I don't think so. I think that all research has to be done, all that's realistically profitable when decided by reasonable men of intelligence. As for the Star Wars speech, I happen to believe that it was pure politics to get the defense budget through Congress.

Arthur C. Clarke, who is credited as the father of the communications satellite, once wrote that every revolutionary idea seems to evoke three stages of reaction which may be summed up by the phrases (1) it's completely impossible; don't waste my time, (2) it's possible, but it's not worth doing, (3) I said it was a good idea all along. We are gathered to talk about the future, or a reasonable facsimile thereof, if there is to be any future. For a generation with so much ahead of it, technologically and exploratively, we seem to have an abnormal fear of the future. We have nothing to fear if we act and act now. And it doesn't apply so much to the engineering types out there as to the public across the country. As Kirkegaard once said, "It is not at all true that the scientist goes after truth; it goes after him." Or, as Wilbur Wright put it back in 1909, "There are three classes of people: one class thinks the flying machine is going to do everything, the second class thinks it's going to do nothing, and the third class gets in the air and sees what it can do."

Too many of us die too young in this country. Medical care is obviously our single most explosive crisis, whether it concerns the ghetto dweller in many major cities, for whom the emergency room has become his doctor's office, or the rural resident, who finds that his doctor's office disappeared a long time ago. A few years ago, everybody was shouting about the doctor shortage. Politicians said we were 50 000 doctors short. Medical schools enlarged their classes, and special loan funds were established. Well, we'll soon have 600 000 practicing doctors in this country, nearly twice the number we had 15 years ago, and it has now become obvious that we never really had a doctor shortage; we had a doctor dislocation, with too many grouped in the most attractive or well-paying cities and too few elsewhere. We still have, by the way, far too many surgeons and far too few GPs, internists, and pediatricians. The most dramatic example I know is that New York City alone has twice the number of neurosurgeons as all of Russia. Medicine in this country is in the middle of a mass care and inflation crisis.

There are 30 million poor in America. For them, death comes earlier, illness is twice as frequent, and there is four times as much chronic illness. The poor believe that poverty is disease and they are right. If you are poor, the risk of dying under the age of 25 is four times the national average.

The really major steps that lie ahead in conquering disease are in brand new fields: bioelectronics, or new kinds of biochemistry that may even eliminate some forms of disease. To get better care, discovery must be initiated and urgently encouraged. New hospitals are needed with more efficient, newer physical plants that

37

can be run more cheaply; otherwise, the nation is on the verge of medically bankrupting itself. The old answers no longer work. New methods of diagnosis that are less invasive, such as CAT scanners or NMR (nuclear magnetic resonance), are part of the answer.

After you've covered as many medical stories as I have and been in and out of as many hospitals as I have, both vertically and horizontally, as newsman and as patient, you realize that the old answers have mostly failed. The cost of medical care has got to be lowered. Some costs, such as labor, are virtually fixed. Roughly 66 percent of every hospital dollar goes for labor. Medical care last year was still the most rapidly rising area of inflation, increasing by 11.7 percent in 1982 and virtually the same percentage in 1981. And it may get worse. The Irving Trust Company in New York ran a study last year of its own medical costs for its employees. The results were frightening. They showed that medical costs rose 40 percent in 1981 and 1982, and their projections for 1983 are another 40 percent. As for medical costs overall, there are some cuts that can be made in lab tests. For example, at costs that now exceed 21 billion dollars a year across the country, too many of these tests are useless or redundant. A University of California study calls them CUTs—"conditionally useless tests"; what a marvelous acronym that is. When the outcome is predictable by any doctor with the information he already has, many such tests come from fear of malpractice suits. Society is being nickeled, dimed, and dollared to death.

As for artificial organs, we can look back on the Barney Clark story just a few months ago. What did it teach us? What did it mean? Beyond his personal saga of courage, there are these simple facts. No more than 5 percent of heart disease cases suffer from worn out heart muscle, or cardiac myopathy, so no more than 5 percent can be helped by artificial hearts or transplants. Compare that to the simple fact that 85 percent of heart disease patients have hardening of the arteries or cardiac artery disease. Those are the ones who need bypasses or perhaps the new laser vaporizing techniques to clean out their plaque, if it works. It does in animals, and many hospitals are now testing it in humans. Another 12 percent have valve disease. So we rapidly see that only a few percent can be helped with artificial hearts. Is it worth the estimated 200 million dollars in research funds that the mechanical heart development costs, in this era of shrinking research funding? I pose the question, but I do not answer it, for I am now wise enough to know that if it's your loved one who is dying and might be helped, no price tag can be put on survival. And although the Barney Clark case may just be the beginning, it is new technology, and as you heard, there is no telling where it can lead.

Let me point out a few things that need more attention and are not getting it. First, not enough young medical scientists are going into research. You can hardly blame them when a third-year resident makes about $25 000 a year, whereas research grants for individuals with the same training average about $16 000 a year. Soon it will not be just autos, steel, and our high-technology industries that are slipping away from us, but medical progress as well. US science has dominated the Nobel prizes for the last decade and longer, but I submit that we are probably living off the past and ignoring the future. Japan doesn't even have such a degree as an MBA, and it's perhaps significant to note that its giant industries, its Sonys and Toyotas, are headed by engineers, not MBAs.

Bright young physicians are drawn into private practice. Unless and until research grants offer competitive living wages, the number of MBAs our universities turn out compared with scientists and engineers is irrelevant. Lawyers, for example, now start at $30 000 to $35 000 a year. Well, you know what scientists start at: $10 000 to $15 000 a year. A promising young scientist can look at a former classmate who's making an arm and a leg in private practice. The young scientist's lot at most universities is to struggle to make ends meet. These young scientists and researchers urgently need to be encouraged. The Achilles' heel of US science is training its scientists and starting them in challenging research jobs.

Despite research grants from the government, we are not making enough progress in the conquest of the basic diseases—heart disease, lung cancer, brain tumors, and asbestosis, or occupational asthma—and in finding the basic pathways by which these diseases work and cause chemical injury to the lungs, the airways, the brain, and other vital sections of the body. We face the enormous task of creating a preventive medicine system for a nation that now practices and always has practiced corrective medicine. This task really overshadows anything America has ever done, from creating nuclear energy to getting to the Moon, because it involves changing the system, telling doctors how and where they can practice, up to a point, and then finding a way to pay for all of it. Believe me, I know; I come from a family of doctors. Yet once this is under way, it will actually save large sums of money by stopping many diseases when they are preventable or correctable, and not yet fatal.

Deep in our crisis (and I see this all around the country, mostly in ghetto cities) is the individual's fear that no one cares, that he has lost his identity as well as his power to do anything about what is happening to him. He feels hopelessly trapped in an ocean of polluted air, rampant inflation with a recession, jammed roads with an energy shortage, run-down housing and broken promises from our politicians—promises that now have to be kept. Our people feel that they're in a rip-off society, and too often they are. The great hope is that we may have just discovered all this in time. In many ways this is because of the news media—the same media, especially television, which is accused of distorting and overplaying the news, and sometimes does so by mediocrity, accident, or deadline, but seldom by intention. That same media may just have saved us by focusing attention on these crises in time, before they get out of control.

You've heard a lot about EPA and dioxin recently; well, let me tell you what's coming up. I happen to know because I've been researching this. You've read about the dioxin sites in Missouri; well, it's likely that there are thousands of dioxin sites across this country, most of them unknown. Dioxin is nothing new, by the way. It was identified as a contaminant left over from manufacturing pesticides, including PCBs and Agent Orange, decades ago. What we don't know about are the long-range effects on our children and their children. Scientists are fearful that it may be a delayed time bomb, like asbestos, that lurks beneath the flesh, only to spring up 15 to 25 years later. The dioxin horror is a national disgrace.

Some days I get so mad that I feel like the speech writer told by his boss that he'd better deliver a flawless job this time or else. The speech writer, intimidated by this, turned out an absolute gem, with the boss telling how he was going to reshape his industry and reduce costs at the same time. After describing several of the steps, the boss came to the end of the page and turned it over; it was completely blank except for the words "OK turkey, you're on your own."

40

One of my favorite slogans is that people who leap to conclusions generally jump over the facts. Well, we'd leap from what is to what could be without stopping in between to be accurate. There are no easy answers to our problems. They'll be costly and take time. We don't even have really good data as yet on how severe many of them are, much less how to conquer them. But technology is the answer—not instant miracles, but the technological excellence that comes from space and aeronautical research, not even to mention the jobs it supplies. Using research in space as a scapegoat may make good politics, but it solves nothing.

As I said earlier, we cannot suppress change; we can and will manage it. As a nation we've got to stop finding reasons why things can't be done and find the reasons why they must be done, as well as the right way to do them, before we wipe ourselves out. The widening gap between what we know and what we do with it, between finite knowledge and infinite failure to use it, threatens to destroy the belief of all of us in our society. As usual, we have more questions than answers, and answers for which there are no questions. But it is glaringly clear that no one is really in charge in Washington to harness and use science and technology; they are playthings of political whim and expediency. No one really has the power and skill, much less the authority, to do the job now needed.

Anyone who thinks that things will ever again be the way they were, or that they should be that way, is dreaming. Not only will this not happen, it should not. Our lifestyle and ways have been challenged and found lacking. What is happening is very much as if we were in the last 15 seconds of the decline and fall of the Roman Empire, with the seconds ticking away, and then the clock was stopped for us for an instant replay and a second chance, a chance to discover how we should go, not only in the decades immediately ahead, but in the third century of our country. Socrates, you'll remember, asked all the important questions but never answered any of them. This generation has to answer those questions, and we're the people who can do it and have to do it. You can call this the age of overact and underthink, or whatever you like. The point is simple: people must prevail. If we lose track of that, we will lose ourselves. So the study of mankind, peoplekind (bioecology, I call it), the art and science of the individual, must be given far more attention than it now receives. Why we do what we do is perhaps far more important than what we do.

The glib words of years past from our politicians are hollow nightmares indeed when we are confronted with the staggering realities of what has to be done. But the key is there—technology, using it—and we hardly do now. The future may be unpredictable, but we can make a few well-aimed guesses about what life will be like in the year 2000. We'll fly on supersonic transports, or more likely hypersonic transports, for which the ground work (or should I call it air work) has already been laid at places like NASA Langley Research Center. We'll have put the genetic code to work and begun to engineer out congenital birth defects and inherited diseases. We'll have made a real start toward the prevention and conquest of both heart disease and cancer. The remaining villains, of course, will be the common cold and hay fever.

The normal life span should be 85 to 90 years, and anybody who suggests mandatory retirement at age 65 will face unthinkable punishment. What that longer lifespan will make possible is two or more careers for most of us. At age 40 or 50, we might go back to school and retrain ourselves for another career. Just think of the possibilities—politicians could even become statesmen. But those promises are distilled from much research by top thinkers, and you've got to watch researchers carefully. Our real goal in research, as one friend persists in telling me, is to reduce utter chaos to mere disorder. Well, we're a little like Columbus before he set sail. When he departed he didn't know where he was going. When he got there he didn't know where he was. When he returned he didn't know where he'd been, and he borrowed all the money to do it with. As George Bernard Shaw once said, "Some men see things as they are and say why; I dream things that never were and say why not." Let me end on a quote from T. S. Eliot, who once wrote, "We shall never cease our exploration, and at the end of our exploring we will get back to where we started and know it for the first time."

And now I'll be happy to tackle any questions you have.

Question: Why do you think there has been the tendency towards a decline in science and engineering education in this country during the last 5 years?

Answer: I think it's the almighty dollar; I think people go where the money is. I know; I have two kids, and they're both trying to make up their minds what to do. A neighbor's son just graduated from Harvard with his MBA and started in at a $35 000 to $40 000 a year job at age 27. Do you know any engineers who start with that kind of money? The only other people I know who start at anywhere near that kind of money are lawyers and doctors. There's something wrong. There's something wormy in the wood of the US infrastructure. There's something wrong in where we place our values. I submit I am terribly worried that the same work that put NACA wings, designed here at Langley, on 707's and Piper Cubs is not being done now, and that when you guys out there retire, the new generation of engineers and aerodynamicists won't be in the works to take over.

Question: Do you have any thoughts or suggestions on how we as technical people might better communicate or interface this upcoming technology with the nontechnical person?

Answer: Yes; explain it! Don't fall back on acronyms. My job is really as a translator between the scientists and the public. If I talk about HSTs or SSTs, to use two examples, or TDRSSs or IUSs, I get my head blasted off by my producers, and for good reason, if I'm assuming that the public understands. Unless they're co-workers or your family, you can't assume anybody understands. So you have to explain what "L over D" is, for example, what a chord is, what a root is, what a dihedral is, what "BLC" is. And then, having done that, you have to go for the gut. You have to say why you think what you're doing is most important or why you think it's not important, or why Langley's work or your section's work is or is not important, and what is important. And don't stop with the public. Go to your congressman, your state legislators.

Question: Given the aviation environment of today, do you see us, without changing the way we're going now, maintaining a superiority as far as aviation technology is concerned, or are we losing that one too?

Answer: I think we're losing that one too. Bill Magruder, the L-1011 project chief for Lockheed, used to say in every speech that aviation was the only favorable balance-of-trade item that we had, or one of the two or three: aviation, agriculture, and electronics. Well, electronics is sure gone, and aviation can't be very far from gone. I suppose some of the foreign carriers are buying Boeing 767's and 757's, but as for the next generation of transports, what company do you suppose is going to gamble on that?

Question: You've covered our first 25 years in space as far as the public is concerned. What do you see for the next 25 years? If someone—one of your friends in Washington—were to ask you what thrust NASA should take over the next 25 years in planetary exploration, Earth observation, and communications, what do you see?

Answer: I would say two things: Mars and the manned space station. We've got to settle the Mars thing. There has to be a companion planet to flee to when some nut starts World War III, because there'll be no place to hide. Well, let's hope that never happens. But that's one of the things I worry about.

Getting back to your original point, I feel that we have done the basic outer-planetary job with Viking, Voyager, and the Pioneers. Now, I do think there is some use for other outer planetary experiments that can be repeated. But I think probably the key thing is the search for life in our solar system, if there is any, and I think it's going to take another variety of unmanned lander on Mars to do that. Why isn't NASA planning that? The second thing is the manned space station, which is just as desperately needed. And the third thing is a new generation of safe small airplanes. A new generation of small planes is a much smaller job in magnitude and dollars than a manned space station or Mars, but it's just as desperately needed.

Question: You pointed to the personal value system as being a problem now and in the future. Well, how much does the media take something like that into account in reporting the news, as far as trying to change personal value systems?

Answer: By personal value system you mean, "How do I make a good living fastest and with the least effort?" I think the media is fairly clear and clean on that one. I don't think we ever get into it, beyond the involvement you just heard from me, and I'm not reporting today. I get very worked up over things like that, when I see MBAs earning what young researchers and young scientists should be earning for essentially sitting around pushing a pencil. I don't think the media really cares, if that's what you're getting at.

Question: What do you see as a specific role that the media can play in correcting the antitechnology attitude that society has?

Answer: I think that I, for one, know I am doing what I can. We should more clearly spotlight the truth as well as the technical failures of our time, and there have been several. The media does the job of clarifying. It forces every public official to be more responsible and to check out projects more thoroughly before investing taxpayers' dollars. That is the role of the media, and I think to a large extent it's being done. In fact, it may be overdone.

Question: I want to ask you a question about the weapons community. It seems that we have a lot of weapons now and we're asking for more. Because of Russia, I don't think we can throw them away, but it seems that we come closer and closer to ending it all by pushing a button. Could you give me your opinion as to the direction Americans could take? Could you include some specifics with regard to the MX, deployment of missiles in Europe, and the B1?

Answer: Let me merely say, to answer your question briefly, that the more equally armed we and the Russians are, the greater the chances of peace, because each nation is fearful of the other. I don't think there's a ghost of a chance of major war or even a minor war ever erupting (a nuclear war, that is) between us and Russia. What I worry about are the maniacs, the smaller powers who are trying to attract attention to themselves. As for the MX, you've seen the Washington confusion about which way to go, and I think it has a distance to go before it straightens itself out.

Question: We read often about the medical, technological, and military research from overseas—say Japan or Russia. Speaking particularly of medicine, do you see in the future the possibility for greater foreign cooperation? I know there has supposedly been some tremendous research on cancer in Sweden, Norway, and England, which we can't take advantage of because of FDA restrictions. I'm thinking, for example, of the stories we've been hearing of the boy, Todd Cantrell, who had the eye problems and went to Russia for treatment.

Answer: It didn't do him any good; in fact, it set him back, according to one of the top eye surgeons. So I urge upon you a greater respect for ambiguity, a greater caution and a greater respect for US medicine. There is nothing in our country which doesn't get approved, although it sometimes takes too long. The FDA is slow to move, but there is no cancer cure that's been held back.

Question: But do you think that countries will ever be able to cooperate in medical research?

Answer: They cooperate right now. One area that we and the Russians still cooperate in (or one of the few areas) is biomedical research. Anybody in the medical arm of the government can tell you that. With the Chinese, too, it's the one area we cooperate with them in. They, for example, have availed themselves of our latest surgical techniques, and we've availed ourselves of what look like very promising herbal cures. You can't beat experience. In truth it's a mix of everybody's cures and everybody's components.

Question: Regarding this boy who went to Russia recently, the media at the time really made every effort to make a story out of it. Everybody in the United States had some opinion about that story at the time because it was blown up by the media. Isn't it a fact that the media to a large degree abdicates its responsibility to our society in the quest of a story for the sake of the story, so that they can get the readership and the television viewership? The media seems to be money oriented; it doesn't seem to want to contribute to society.

Answer: You're right. The media did blow the Todd Cantrell story completely out of proportion. However, I can also throw into that mix the hour-long asbestos documentary I did last December, when I threw the book at both industry and government for failing to protect us when they had known about the problem for 30 or 40 years. But don't tell me you haven't heard me speak of the media's failings, because there are many. There are too many Todd Cantrell stories.

Question: Is there any sense of responsibility to society, not on an individual case, but on a media-wide base, or is it just that the companies have to answer to their stockholders?

Answer: That doesn't get into it at all, because commercial minutes on the news shows on all three networks sell for about the same amount of money. Each rating point is a couple of thousand dollars a week, but the news shows all lose catastrophic amounts of money; they're loss leaders, if you will. So even if we can raise our ratings and our share of points over CBS or NBC, that's not a governing factor. There is only one governing factor, and that is attracting the audience with the truth. That's my mission at ABC, but sometimes I'm not there, or I'm off on other projects. I was off on another project the day the Todd Cantrell story ran. The next day I asked an eye surgeon in New York, "Can this Russian treatment help this kid?" She said, "No." I tried to make the point, but it was too late, the story was gone. We did make the point when the kid came back, although by then the parents were elated because their son was potentially cured. People in that role are searching for what I call magic helpers. They're not searching for the truth—they don't care about the truth—they're looking for magic helpers.

Question: May I pursue one more point? There are, on your network, programs that in my opinion masquerade as news programs. These hour-long programs come out with more or less sensationalist stories, and they serve the same purpose as the *National Enquirer*, as far as I'm concerned, because the sensationalism is there to generate points. I don't think we had as many of these a few years ago. The documentaries that were of value have all disappeared.

Answer: All I can tell you is that ABC does 12 documentaries a year, more than the other two networks put together. We're the only active documentary unit in the country among the major networks.

Question: You noted that Socrates asked political and world questions that are still important and pertinent today, and you also noted that science has advanced immensely since Socrates' time. Does this mean you agree that technology is advancing faster than our political institutions?

Answer: I didn't say that. I said Socrates asked all the important questions, but never answered any of them. It is for our generation to answer them.

Question: Could you give us some suggestions on how to avoid problems that might result from this uneven growth between science and the political institutions?

Answer: If I knew that I wouldn't be a mere science editor.

Question: I have a question and a statement. If we have to answer the questions Socrates asked, we're in trouble. As for my question, I'm very interested in artificial intelligence and the development of robotics, including the consequences of this development for our political and social institutions. Two interesting theories about the development of robotics are that it might revamp the scarcity-based theories on economics and change our economic system, and it might add a certain bundle of commodities or basic services to our political rights and what we consider to have inherited from our economic and political institutions. Could you just comment on what you think about the evolution of robotics and artificial intelligence, politically and economically, over the next 20 years?

Answer: Well, I would point out to you the conclusions reached by the recent conference of top business leaders in the country, including those in the steel and auto industries. They pointed out that many of their workers who have been laid off will never be rehired because their jobs have been either taken over by automation or eliminated. Obviously, it's a major problem that we have to deal with. You can have a robot science editor too, or a robot to do your job. It won't be as skillful, perhaps, but it will be cheaper.

Question: You spoke several times of the international exchange of ideas and scientific knowledge, particularly the biomedical exchanges that even this country and the Soviet Union can be happy with. Do you think there's a possibility that over the next 20 years this type of scientific interchange could help bring about more of a global perspective on citizenship, which would then have some positive effects on the political conflicts that are present right now? Could such an interchange give people a more global concept of society to replace the nationalism that is present now?

Answer: Good question. Let's say the Russians come up with a cure for one kind of cancer and we come up with a cure for another kind of cancer; very quickly you're going to see an international exchange. Now if you want to extrapolate that (and there are obvious dangers, as I pointed out, in extrapolation) to future space activities, the next great exploration in space has to be Mars. The studies we have (and there are only a few) indicate that a manned Mars expedition will be very expensive. A study done in 1970 for the Air Force estimated that with a six-man crew, a two-year round-trip flight would cost $100 billion in 1970 dollars. Carry that forward to today and it would cost about $350 billion. By the time such a project gets started it will probably cost $500 billion, which is more than any single country can afford. You see what I'm getting at—it will have to become an international project. There is great hope for unifying the world in that way, and for eliminating hostilities.

Question: So you see the space exploration system as the way to break down some of the conflicts?

Answer: Yes.

Question: A few years ago I heard one of our congressmen talk about using the Sun's energy by collecting it with a dish in space, changing it to microwave energy, transporting it to Earth like a radar beam, and then converting it back to use as energy. Do you have any knowledge of this program?

Answer: The congressman was talking about what is called the solar power station. It's still being studied by NASA and the Department of Energy, but there are several problems as well as much promise. One problem is that if you're a half degree off in aiming the microwave energy to Earth, instead of beaming it to the five-mile-square field outside Phoenix you might just burn up Phoenix. So it has to be totally reliable. What's more, you have to put a lot of hardware into orbit, requiring a much larger boosting system than the Shuttle has.

Question: Can you give me an estimate on when you think this could be accomplished?

Answer: At the rate we're going now?

Question: Yes; we had been investigating this for a few years and then all of a sudden it was just left hanging.

Answer: All of a sudden the energy crisis slowed down and gasoline went below a dollar a gallon, so the project was put on the back burner.

Question: I have a question about the Space Telescope. How much coverage will there be of the launch, and after the Space Telescope is in orbit and operating, how much coverage will be available?

Answer: You're talking about 3 years or so from now, and there are new reports of major problems in both the mirrors and the optics of the Space Telescope, which had been scheduled for launch in 1986. It will have live TV cameras on it; all three networks will bring you the pictures.

Question: Concerning propulsion as a source of energy, how far along are we coming in the area of nuclear fission?

Answer: You mean nuclear fusion. Not as far as we should, but we're getting there slowly.

Question: I think it's obvious to all of us that the Three Mile Island incident had a disastrous effect on the nuclear industry. That, coupled with lagging electrical sales demands, makes me fairly confident that no new nuclear plants will come on line in this country between now and the year 2000. With the existing nuclear power plants producing electricity on a commercial basis, what is it going to take to get the Federal government to take a decisive stance on the problem of storage of the nuclear waste that we're producing at present?

Answer: I don't think you understand fully the fact that nuclear power plants produce only low-level radioactive waste. The high-level radioactive waste is from weapons systems, and they're the more immediate problem.

Question: There is an immediate problem in the sense that some of these nuclear power plants are going to be forced to shut down within the next year or two because they don't have the storage available.

Answer: Well, we're talking politics again. I don't pretend to be any great nuclear expert. It's tough enough just keeping up with space and aviation. One of the problems we have is that our whole society has gotten so complex that no one person can keep up with anything. All these reprocessing plants that could have handled much of the waste are still mired in political battling. So go see your congressman.

Question: What's it going to take to get us out of that political mire? Do you suspect the government will actually take a decisive stance, or can the media help with this problem?

Answer: Probably it's going to take a new Arab oil embargo to bring us to our feet, begging for oil and thus begging for nuclear power. That's the only real chance of getting out of it. The media, as you have noticed, is antinuclear, except for iconoclasts like me who perversely persist in telling the truth.

Question: I work for the power company, so I happen to agree with that. My last question, if you don't mind, is how serious do you think this dioxin problem is?

Answer: We don't know. We don't have enough data on it, and the fact that we don't is just downright disgraceful because dioxin is nothing new and we should have had the data by now. All we know is that dioxin causes chloracne, which goes away, and also some neurological disorders. The Swedes claim it causes bone cancer, but some of our people dispute that study. I have yet to see a definitive study on it.

Question: You mentioned the lack of understanding and appreciation of science in Washington. Do you have any idea how this situation arose and why it continues today? Is there any relation between the lack of understanding and appreciation of science in Washington and the fact that, as you mentioned, Japanese companies are usually headed by engineers instead of MBAs?

Answer: No, I don't think that's related to the lack of appreciation of science in Washington. I think you're talking about apples and oranges. I don't know how or why Washington became disenthused with science, but I suspect that it was a few too many Three Mile Islands, a few too many power companies lying and failing to do their job, a few too many regulatory agencies not doing their job to protect the public, as is implicit in their charter.

51

Question: You said that we will be technologically illiterate in the future. Do you think this is inevitable, or can we prevent it?

Answer: That's what the National Academy of Sciences warns. There are steps already under way to upgrade science and math education in high schools, but it's going to be a long time before the problem is solved. For example, you can put in computers for 6-year-olds to use, but who is going to do the software programming and teach the teachers how to use them?

Question: In light of dioxin, Agent Orange, and everything else we read about in the newspapers, do you see a more conservative trend in research in that area or are we just going to be reading more and more about it in the next 30 years or so?

Answer: Do I see a more conservative trend in chemical research on pesticides? I certainly hope not. I think what's needed is a more conservative system of checks and controls but not necessarily a more conservative trend in research. I think this country's great advances have been made because we took on the impossible and proved it possible, but when something is basically dangerous, like dioxin, and people know it, that's something else again.

Question: I'm an engineer, and I remember when you talked about the difference between what you earn in basic research and what you earn out in the marketplace working in a production-type atmosphere. Very little basic research is done without government sponsorship. Do you think that private industry should be taking on more of this responsibility in order to promote a higher standard of living for people doing basic research?

Answer: Yes.

Question: I would like to know what you think about the advancement of research in new materials. Do you think this research has reached a peak as far as the resources we have on Earth, or do you think we'll be able to go to other places, such as the Moon, for new materials?

Answer: Do I think the advancement of research in new materials has reached a peak on Earth? Well, you could have said that 5 or 10 years ago, and then along came Kevlar or some other new material. Every time we say we've reached the limit, there's no further to go, this is the peak, some quiet little guy in a a lab comes up with something new. As for the Moon, I know of no chemical or mineral found on the Moon that is any different from those found on the Earth. The proportions vary, but when you factor in the cost of getting material back or building a lunar camp to refine it, it comes out as being more expensive.

Question: A few years ago, during the energy crisis, there were plans for plants to convert coal to oil. Now they've discarded the whole idea, although I understand it was feasible. Can you tell me why you think it was discarded, besides the fact that the energy crisis is over, and whether you think it will ever be renewed or should be renewed?

Answer: I think you've answered your own question in your statement. This country has, as you've read in all the ads, 80 percent of the world's coal. Five or ten years ago, at the height of the energy crisis, one top scientist said, "The best way to gasify coal is to burn it."

Question: You mentioned earlier that Russia, Japan, and China graduate many more engineers than the US does. How much of the cost of that education was handed down to the individual student, compared to in the United States?

Answer: I don't know, but obviously the state assumes a very high percentage of it, I would think.

Question: Wouldn't you say that would account for our lack of engineers?

Answer: No, I don't think so at all. I think that if others were given the freedom of choice that we have in this country, the freedom to commit suicide, if you will, the freedom to sit back as a big fat-cat MBA and rake in the dollars, they would do it too. But I think in those countries the state decrees, "You will be an engineer, you will be a scientist," and then offers attractive lures.

Question: What I'm trying to say is that I would much rather be an electrical engineer right now, but my pocketbook says I'm going to have to settle for being an electronics technician.

Answer: Join the parade.

Question: How do you see the tremendous increase in technology affecting the Third World countries?

53

Answer: Favorably, I trust, so they become relatively faster second-world countries and then first-world countries, so it doesn't take them three centuries or two centuries or even a century and a half to get to our point.

Question: Do you think there is a reason for us to share this technology with them or is there more of a reason for us to keep it to ourselves and keep them where they are?

Answer: I know of no technology except highly classified technology, which the Russians seem to steal anyway, that this country hasn't openly shared with the rest of the world.

Question: Referring to the deficiencies in our education system which you spoke of, would you attribute them to a lack of impetus on the part of the student or a lack of incentive on the part of the schools?

Answer: Both.

Question: When research projects such as the ion engine, coal gasification, or solar power go stale, is that because of the government falling back, saying we don't need it, or the scientist saying we don't want to do it?

Answer: I think it's a combination of both. I think if the scientists say, "We don't want to do it," the government then says, "It's not needed."

Question: On the subject of space exploration, what types of fuels are presently available and what types are being studied?

Answer: The available fuels are primarily the liquid and solid fuels, but everything from solar sails to nuclear propulsion is being studied. I've contended for some time that we can't get to Mars, for example, with a manned flight (presently a 2-year mission, round trip) using the conventional chemical rockets, even liquid hydrogen and liquid oxygen, unless we have more efficient, higher thrust engines. The speed required to escape the Earth's gravitational pull is about 25 000 mph, or about 35 000 feet per second. But then, as many of you know, at midpoint the vehicle will slow down to 4000 mph until the next planet's gravitational attraction begins to pull it in. That's where you've got to make up the time with some sort of constant drive; it could even be an electron motor. They're all being studied.

Question: How do you think the advancement of technology is going to affect the environment that we all share, and do you think the technology that is proposed and discovered is going to be able to keep up with the problem of storing our nuclear waste in a way that will not affect future generations?

Answer: That's a sweeping question you've asked, but I can answer it by simply saying that the gross crimes of the past two decades have guaranteed that out of the ashes of the old EPA will come a new EPA. I believe that no new technology that pollutes will be allowed.

Our Future in the Cosmos—Computers

Isaac Asimov

Isaac Asimov

For more than 45 years, Isaac Asimov has been a professional writer of renowned versatility. The Russian-born scientist and author, who has been called a genius and "the nearest thing to a human writing machine," is perhaps best known as one of today's major science fiction writers. His broad range of works includes histories, childrens' books, collections of articles, mysteries, and books concerning the Bible, literature, geography, and nonfiction science material.

Blessed with total recall, Asimov entered Columbia University at the age of 15 and graduated with a doctorate in chemistry. Beginning in 1949, as instructor, associate professor, and later full professor, he taught biochemistry at Boston University's School of Medicine. His scientific research includes work in kinetics, photochemistry, enzymology, and irradiation.

Asimov's impressive list of writings includes hundreds of articles in publications ranging from *Esquire, Harpers,* and *The Saturday Review,* to pamphlets of the Atomic Energy Commission. His latest and just-published novel, *Robots of Dawn,* is the third book in a series concerning his fictional character, Elijah Baley, and he is currently working on a revision to *Asimov's Guide to Science* and on a new novel, *Opus 300,* which will be his 300th book. He is the recipient of numerous awards, including the American Association for the Advancement of Science Westinghouse Award for excellence in magazine writing and a 1983 Hugo Award from the World Science Fiction Convention for his novel *Foundation's Edge,* a sequel to a trilogy of novels that he wrote 3 decades ago concerning the distant future.

Known to type 90 words a minute and to produce as much as 35 pages of manuscript a day, Asimov maintains an 8-hour-a-day, 4-day-a-week writing schedule and calls writing "my idea of a vacation. Most of all, I want to be writing," he says. "If I could, I'd write every book in the world." Dr. Asimov lectures as enthusiastically as he writes, and has been referred to as one of the "great explainers" of our technological age, helping to bridge the gap between science and the public.

Our Future in the Cosmos—Computers

*No matter how clever or artificially intelligent
computers get, and no matter how much they help us
advance, they will always be strictly machines and we
will be strictly humans. When we finally do extend
the living range of humanity throughout near space,
possibly throughout the entire solar system and out
to the stars, it will be done in tandem with advanced
computers that will be as intelligent as we are, but
never intelligent in the same way that humans are.
They will need us as much as we will need them.*

As far as our destiny in the cosmos is concerned, I think that it
will arise out of the two important changes that are taking place
before our very eyes. One change involves the computerization
of our society, and the other change involves the extension of our
capabilities through aeronautical and space research. And the two
are combined. Decades ago we science fiction writers foresaw a
great many things about space travel, but two things we did not
foresee. In all the time that I wrote stories about our first Moon
landing and about the coming of television, nobody, as far as I
know, in the pages of the science fiction magazines, combined the
two. Nobody foresaw that when the first Moon landing took place,
people on Earth would watch it on television. Nor did science
fiction writers foresee that in taking ships out into space, they
would depend quite so much on computers. The computerization
of space flight was something that eluded them completely. So, I
have two broad areas that I can discuss in talking about our destiny
in the cosmos. One area is the future of computerization, and the
other area is the future of space itself. In this presentation, I will
talk about computers and their future, and I think I have a kind
of right to do so. I have never done any work on computers, but I
have speculated freely concerning them.

Despite my gentle appearance as a gentleman a little over 30, I
have been a published writer for 45 years. If I can make it 5 more
years, I will celebrate my golden anniversary as a published writer,
which isn't too bad for a fellow in his early thirties. Perhaps

the most important thing I did as a speculator was to foresee the various properties and abilities of computers, including those mobile computerized objects called robots. As a matter of fact, I sometimes astonish myself. Back in 1950, in a passage that was eventually published as the first section of my book *Foundation*, I had my protagonist pull out a pocket computer. I didn't call it a pocket computer, I called it a "tabulator pad." However, I described it pretty accurately, and this at a time when computers filled up entire walls! Decades later, someone said to me, "Hey Asimov, you described a pocket computer a long, long time ago; why didn't you patent it and become a trillionaire?" And I said, "Did you notice, perchance, that I only described the outside?" I'll be frank, to this day I don't know what is inside. I have evolved a theory; I think it's a very clever cockroach. But that was in 1950; I did a lot better in 1939, long before many of you were born. I began writing about robots. Robots had been written about for years before this. The word had been invented in 1921 by Karel Capek, a Czechoslovakian playwright. However, until I started writing, robots were, for the most part, either menaces or sort of wistful little creatures. As menaces, they always destroyed their creator; they were examples to humanity of what to avoid. They were symbols of the egregious hubris of the scientist. According to that plot you did something that infringed upon the abilities reserved for the creator, you made life. No one objected to destroying life, you understand (don't let me get radical here), but making life was wrong, especially if you didn't use the ordinary method. Even if you did use the ordinary method, the robot, as though to explain to you that you had done wrong, sometimes killed you. Well, I got tired of that plot. There was another plot in which the robot was a good and noble but picked-on member of a minority group but everyone was mean to him. I got tired of that plot too. I decided that robots really ought to be (hold your breath now) ... machines, to do work that they were designed to do, but with safeguards built into them. Looking around the world, I noticed that practically everything human beings had made had elements of danger in them, and that, as best we could (being fallible human beings), we had safeguards built in. For instance, you will notice that swords have hilts, so that when you thrust the sword forward and it goes into the other guy the way it is supposed to, your hand doesn't slide along the blade and cut all your fingers off. So, I figured robots would also have safeguards built in, and I finally listed these safeguards in the March 1942 issue of *Astounding Science Fiction* on page 100, first column, about one-third of the way down. Since then, I have had occasion to look up the list and memorize it. I called it the "Three Laws of Robotics." I will now recite these laws for you because I have memorized them. I have made a great deal of money from them, so it sort of warms my heart to think of them for purely idealistic reasons.

1. A robot may not injure a human being, or, through inaction, allow a human to come to harm.

2. A robot must obey orders given to him by human beings except where such orders would conflict with the First Law.

3. A robot must protect its own existence as long as such protection does not conflict with the First or Second Law.

None of these laws is interesting in itself, although it is obvious that the laws apply to all tools. If you stop to think, the first rule of any tool is that you operate it safely. Any tool that is going to kill you when you use it is not going to be used. It won't even be used if it merely maims you! The second rule is that a tool should do what it is supposed to as long as it does so safely. And the third rule is that a tool ought to survive its use and be ready for a second use, if that can possibly be arranged. Nowadays, people who are working with robots actually debate the methods by which these three rules can be installed. This flatters me, but what interests me most is that I called these rules the Three Laws of Robotics, and that use of the word "robotics" in the March 1942 issue of *Astounding Science Fiction*, (page 100, first column, one-third of the way down) was the first use of this word anywhere in the English language. I made up the word myself; this is my contribution to science. Someday, when a truly encyclopedic history of science is written (you know, one with 275 volumes), somewhere in volume 237, where the science of robotics is discussed, there will be a footnote: The word was invented by Isaac Asimov. That is going to be my only mention in all 275 volumes. But, you know, better than nothing, I always say. The truth is that I didn't know I was inventing a word; I thought it was the word. If you will notice, physics ends in "ics" and just about every branch of physics, such as hydraulics, celestial mechanics, and so on, ends in "ics." So I figured that the study of robots would be robotics, and anyone else would have thought of that too if they had stopped to think that there might be a study of robots. What's more, I quoted those three laws from the *Handbook of Robotics*, 56th edition (c.2058 A.D.), and the first edition of such a handbook is actually about to come out. It is a handbook of industrial robotics, and I was asked to write the introduction. Who would have thought, when I was a little kid writing about robots, that such a handbook would actually be written! It just shows that if you live long enough, almost anything can happen.

The question then is: What is going to happen with robotics in the future? Well, as we all know, it's going to create a certain amount of economic dislocation. Jobs will disappear as industries become robotized. What's more, robots are dangerous, very literally dangerous sometimes. There has already been one case of a robot killing a human being. A few years ago a robot in a Japanese assembly line stopped working, and a young mechanic went to see what was wrong with it. The robot was surrounded by a chain-link fence, and the safeguard system was designed to cut off power to the assembly line when the fence gate was opened, thus deactivating the robot and making it just a lump of dead metal. This safeguard was designed to implement the First Law: Thou shalt open the chain fence before you approach the robot. (You have to understand that what we call industrial robots are just a bunch of computerized levers, nothing more. They're not complicated enough to have the three laws built into them, so the laws are implemented outside them.) Well, this mechanic thought he would save himself a second and a half, so he lightly jumped over the fence and manually turned off that particular robot. This will do the trick just as well, unless you happen to push the "on" button with your elbow while you're busy working on the robot. That is apparently what he did, so the robot, in all innocence, started working. I believe it was a gear-grinding robot, so it ground a gear in the place where a gear was supposed to be, which was where the guy's back really was, and it killed him. The Japanese government tried to keep it quiet, because they didn't want anything to spoil their exploitation of robotics. But it is difficult to keep a thing like that completely quiet. Eventually the news got out. In all the newspapers the headlines were: "Robot Kills Human Being."

When you read the article you got the vision of this monstrous machine with shambling arms, machine oil dribbling down the side, sort of isolating the poor guy in a corner, and then rending him limb from limb. That was not the true story, but I started getting telephone calls from all over the United States from reporters saying, "Have you heard about the robot that killed the human being? What happened to the First Law?" That was flattering, but I suddenly had the horrible notion that I was going to be held responsible for every robotics accident that ever happened, and that made me very nervous. I am hoping that this sort of thing doesn't happen very often. But the question is: What's going to happen as robots take over and people are put out of jobs? I am hoping that that is only a transition period and that we are going to end up with a new generation that will be educated in a different way and that will be ready for a computerized world with considerably more leisure and with new kinds of jobs. It is the experience of humanity that advances in technology create more jobs than they destroy. But they are different kinds of jobs, and the jobs that are going to be created in a computerized world are going to require a great deal more sophistication than the jobs they destroy. It is possible that it won't be easy to reeducate or retrain a great many people who have spent their whole lives doing jobs that are repetitive and stultifying and therefore ruin their brains. Society will have to be extremely wise and extremely humane to make sure that there is no unnecessary suffering during this interval. I'm not sure that society is wise enough or humane enough to do this. I hope it is. Regardless, we will eventually come to a period when we will have a world that is adjusted to computerization.

Perhaps then we will have another and even more intractable problem. What happens if we have computers and robots that are ever more capable, that are ever more versatile, and that approach human activity more and more closely? Are we going to be equaled? Are we going to be surpassed? Will the computer take over and leave us far behind? There are several possible answers to these questions, depending upon your mood. If you are in a cynical mood, if you have been reading the newspapers too closely, the answer would be: who cares? Or if you have become even more cynical, the answer would be: why not? You might look at it this way: The history of humanity is a long tale of misery and cruelty, of destroying each other and the Earth we live on, and we don't deserve to continue anymore. If there is anything with more wisdom than people, with better brains than we have, that can think better ... please let it take over. You might also argue this way: For 3½ billion years, life has been evolving on Earth very slowly, in a hit or miss fashion, with no guiding principle, as far as we can tell, except survival. As the environment changes and the Earth undergoes various changes, life takes advantage of new niches, fumbles the old niches, and eventually, after 3½ billion years, finally develops a species with enough brains to create a technological civilization. That is a long time to achieve so little. You must ask yourself, "Is that the best that can be done?" Well, maybe that is the best that can be done without a guiding intelligence, but, after 3½ billion years, the guiding intelligence, such as it is, has been created, and now it can take over. You can even argue that the whole purpose of evolution has been, by hit and miss, to finally create a species that can then proceed to accelerate evolution in a guided direction. In that case, we are designing our own successors. Instead of waiting several million years just to develop enough of a Broca's convolution in our brains so that we can learn to talk, we are deliberately designing computers so that they can speak, understand speech, and do a few more things. If there is a grand designer up above who has chosen this way of creating human beings, he is going to be rubbing his hands and calling for applause. He is going to say, "Watch the next step, this is going to be faster than you can possibly imagine."

On the other hand, there is another way to look at the future of the computer. You can also assume that you are not going to easily manufacture artificial intelligence that will surpass our natural intelligence. Miserable as we are, and deplorable as our records prove us to be, we nevertheless have 3¼ pounds of organic matter inside our skulls that are really worthy of 3½ billion years of evolution. When you are being cynical you can speak of the brain as something so little to take so much time to evolve. But consider it closely, it is extremely astonishing, really! There are 10 billion neurons in the human brain, along with 10 times that many supporting cells. Each of these 10 billion neurons is hooked up synaptically with 50 to 500 other neurons. Each neuron is not just a mere on/off switch; it is an extremely complex system in itself, which contains many sets of very strange and unusual molecules. We really don't know what goes on inside the neurons on an intimate basis, nor do we know exactly the purpose for the various connections in the human brain. We don't know how the brain works in anything but its simplest aspects. Therefore, even if we can make a computer with as many switches as there are human neurons, and even if we can interconnect them as intricately as they are connected in the human brain, will the computer ever be able to do what we can do so easily? Now there are some things in which a computer is far ahead of us. Even the simplest computer, the very first computer, for that matter even before they became totally electronic, was far ahead of us in solving problems and manipulating numbers. There is nothing a computer can do that we can't if we are given enough time and if we correct our errors. But that's the point; we don't have enough time and we don't have the patience or the ability to detect all our errors. That is the advantage of the computer over us. It can manipulate numbers in nanoseconds without error, unless error is introduced by the human beings who give it the instructions. Of course, as humans we always like to think of the other side of this argument. My favorite cartoon, in the *New Yorker*, shows a computer covering an entire huge wall, as they did in those days, and two little computer experts. (You could tell that they were computer experts because they wore white lab coats and had long white beards). One of them is reading a little slip of paper coming out of a slot in the computer. He says, "Good heavens, Fotheringay, do you realize that it would take 100 mathematicians 400 years to make a mistake this big?"

The question is: If computers are so much better than we are in this respect, why shouldn't they be better than we are in all respects? The answer is that we are picking on the wrong thing. This business of fooling around with numbers, of multiplying, dividing, integrating, differentiating, and doing whatever else it is that computers do, is trivial, truly trivial! The reason that computers do so much better in these areas than we do is because our brains are not designed to do anything that trivial. It would be a waste of time. As a matter of fact, it's only because we are forced, in the absence of computers, to do all this trivial work that our brains are ruined. It is like taking an elaborate electronic

65

instrument and because it happens to be hard and heavy using it as a hammer. It may be a very good hammer, but obviously you are going to destroy the instrument. Well, we take our brain and what do we use it for? We file things alphabetically, make lists of things, work out profit and loss, and do a trillion and one other things that are completely trivial. We use our fancy instrument for trivia simply because there is nothing else that can do it. Now enters the computer. The computer is a halfway fancy instrument. It's a lot closer to a hammer than it is to a brain. But it's good enough to be able to do all those nonsense things that we have been wasting our brains on. The question is, what then is it that our brain is designed for? The answer, as far as I'm concerned, is that it is designed to do all sorts of things that involve insight, intuition, fantasy, imagination, creativity, thinking up new things, and putting together old things in completely new ways; in other words, doing the things that human beings, and only human beings, can do.

It is difficult for me to put myself into the minds of others; I can only put myself into my own mind. For example, I know that I write stories, and I write them as fast as I can write. I don't give them much thought because I'm anxious to get them down on paper. I sit and watch them being written on the paper as my fingers dance along the keys of my typewriter, or occasionally of my word processor. I start a story in the right place and each word is followed by another word (a correct other word) and each incident is followed by another incident (a correct other incident). The story ends when it is supposed to end. Now, how do I know that all the words are correct, all the ideas are correct, and all the incidences are correct? I don't know in any absolute sense, but at least I can get them published. I virtually never fail! The thing is, I literally don't give it any thought. People ask me, "How can you write all the stuff that you write?" (I have written 285 books at the moment, and I've been busy writing *Opus 300* so it can be published as my 300th book.) I say, "Well, I cut out the frills, like thinking." Everyone laughs; they think I'm being very funny. But I'm not; I mean it literally. If I had to stop and think, I couldn't possibly do all that I do. All right, I'm not as good as Tolstoy, but considering that I don't think, I'm surprisingly good. Of course, the real answer is that I don't consciously think. Something inside my brain puts the pieces together and turns out the stories. I just don't know how it's done.

It is a similar situation if I want to flex my arm. I don't know how the devil I do it. Some change is taking place in my muscle molecules, in the actomyosin, which causes them to assume another shape. There is a ratchet or something that drags the actin molecules along the myosin; who knows? The theory changes every year. But whatever it is, I say to myself, "flex" and it flexes. I don't even know what I did; in fact, I don't even have to say "flex." If I'm driving my automobile and something appears before me, my foot flexes and stamps down on the brake before I can say to myself "brake." If it didn't do that before I could say to myself "brake," I wouldn't be alive now. The point is that our brain does things, sometimes very complex things, that we don't know how it does. Even the person who does it doesn't know how he does it. If you don't want to take me as an example, consider Mozart who wrote symphonies at the ridiculously early age of 7 or 8. Somebody wrote to him when Mozart was an old man of 26 and asked him how to go about writing symphonies. Mozart said, "I wouldn't if I were you; you are too young. Start with something simple, a concerto or sonata; work your way up to symphonies." The guy wrote back and said, "But Herr Mozart, you were writing symphonies when you were a little boy." Mozart wrote back, "I didn't ask anybody."

It's quite possible that we will never figure out how to make computers as good as the human brain. The human brain is perhaps a little more intractable than we imagined. Even if we could, would we? Is there a point to it? There may not be, you know. We talk about artificial intelligence as though intelligence is a unitarian, monolithic thing. We talk about intelligence quotient as though we can measure intelligence by a single number. You know, I'm 85, you're 86, you're more intelligent than I am. It's not so. There are all sorts of varieties of intelligence. I believe that people who make up intelligence tests make up questions that they can answer. They've got to! Suppose I want to design a test to decide which of you has the potential to become a great punk rock musician. I don't know what to ask; I know nothing about punk rock. I don't even know the vocabulary. All I know are the words "punk rock." So this is not the kind of test I can make up. My point is that we have a whole set of intelligence tests designed by

67

people who know the answers to the questions. You are considered intelligent if you are like they are. If you're not like they are you rate very low. Well, what does that mean? It just means that it is a self-perpetuating process.

I am fortunate—I happen to have exactly one kind of intelligence, the kind that enables me to answer the questions on an intelligence test. In all other human activities I am abysmally stupid. But none of that counts; I'm tabbed as intelligent. For instance, suppose something goes wrong with my car. Whatever it is that goes wrong, I don't know what it is. There is nothing that is so simple about my car that I understand it. So, when my car makes funny sounds, I drive it in fear and trembling to a gas station where an attendant examines it while I wait with bated breath, staring at him with adoration for a god-like man, while he tells me what's wrong and fixes it. Meanwhile, he regards me with the contempt due someone so abysmally unintelligent as to not understand what is going on under the hood. He likes to tell me jokes, and I always laugh very hard because I don't want to do anything to offend him. He always says to me, "Doc," (he always calls me Doc; he thinks it's my first name). "Doc," he says, "A deaf and dumb man goes into a hardware store. He wants nails, so he goes up to the counter and goes like this and they bring him a hammer. He shakes his head and he hammers again. So, they bring him a whole mess of nails. He takes the nails that he wants, pays for them, and walks out." And I nod. Then the attendant says, "Next a blind man comes in and he wants scissors. How does he ask for them?" I gesture to show scissors but the attendant says, "No, he says, 'May I have a pair of scissors?'" Now, from the dead silence I always get when I tell this joke, I can tell that you agree with my answer. But a blind man can talk, *ipso facto*, right? All right! Well that shows your intelligence. It is an intelligence test, right there, and every one of you probably flunked! So, I maintain that there are all kinds of varieties of intelligence and that's a good thing too because we need variety. The point is that a computer may well have a variety that is different from all the human varieties. In fact, we may come up with a whole set of varieties of intelligence. We would have a number of species of the genius-human intelligence and a number of species of the genius-computer intelligence. That's the way it should be; let the computers do what they are designed to do and let the human beings do what they are designed to do. Together, in cooperation, man and computer can advance further than either could separately. Of course, it is possible to imagine that we could somehow design a computer in such a way that it could show human intelligence, have insight and imagination, be creative, and do all the things we think of as typically and truly human. But, so what? Would we build such a computer, even if we could? It might not be cost effective.

Consider it this way—we move by walking. We lift first one leg, then the next one; we are consistently falling and catching ourselves. This is a very good method of locomotion because we can step over obstacles that aren't large, we can walk on uneven roads or through underbrush, and we can make our way through crowds. Other animals move differently; they jump, hop, fly, swim, glide, and so on. Finally, we invented artificial locomotion with the wheel and axle. It's one method that no living creature has developed. There are good reasons for that; it would be very difficult for a living creature to have a wheel and axle supplied with nerves and blood vessels. Nevertheless, we have both artificial locomotion on wheels and human locomotion on legs, and each has its advantages. We can move a lot faster on a machine. On the other hand, when we walk we don't need a paved highway or steel rails. We'd have to make the world very smooth and convenient if we were going to take advantage of wheels. But, it's worth it, at least most of us think so. I've heard no suggestions that we go back to walking to New York. On the other hand, walking is not passé. I frequently have occasion to navigate from the bedroom to the bathroom sometime in the dead of night, and I tell you right now, I'm never going to take an automobile to do that. I'm going to walk; that is the sort of thing walking is for.

The question is, can you invent a machine that will walk? Of course you can! I've seen machines that can walk, but they're usually merely laboratory demonstrations. These machines might have very specialized uses, but I don't think they can ever really take the place of walking. We walk so easily that it makes no sense to kill ourselves working up a machine that will walk. And, as far as computers and human beings are concerned, it is wasteful to develop a computer that can display a human variety of intelligence. We can take an ordinary human being and train him, from childhood on, to have a terrific memory, to remember numbers and partial products, and to work out all kinds of shortcuts in handling addition, subtraction, multiplication, division, square roots, and so on. In fact, people have been born with the ability; they are mathematical wizards who can do this sort of thing from an early age, and sometimes they can't do anything else. But once you train that ordinary human being, what do you have? You have a human being, which you've created, so to speak, at enormous effort and expense, who can do what any cheap two-dollar computer can do. Why bother? In the same way, why go through the trouble of building an enormously complex computer, with complicated programming, so that it can create and write a story when you have any number of unemployed writers who can do it and who were manufactured at zero cost to society in general, by the usual process. To sum it up, I think we can be certain that no matter how clever or artificially intelligent computers get, and no matter how much they help us advance, they will always be strictly computers and we will always be strictly humans. That's the best way, and we humans will get along fine.

The time will come when we will think back on a world without computers and shiver over the loneliness of humanity in those days. How was it possible for human beings to get along without their friends? You will be glad to put your arm around the computer and say, "Hello, friend," because it will be helping you do a great many things you couldn't do without it. It will make possible, I am sure, the true utilization of space for humanity. When we finally do extend the living range of humanity throughout near space, possibly throughout the entire solar system and out to the stars, it will be done in tandem with advanced computers that will be as intelligent as we are, but never identically intelligent to humans. They will need us as much as we need them. There will be two, not one of us. I like that thought.

Question: In your writings, do you address the issues of how far society wants computer technology to develop and how man's preoccupation with his own preservation will inhibit the computer's full utilization?

Answer: Actually, I did address those issues in my writings. The gentleman points out that it's possible human beings may decide that there are limits to how far computers should be allowed to go. It may not be what computers may actually do, but what they may threaten in the human mind, and what humans may think of them. In my robot stories, I used the Frankenstein complex in which human societies refuse to allow robots to work because they have decided they don't want to lose their jobs, they don't want to undergo the painful period of transition, and they don't trust the robots to be harmless. They call a halt to computer development. The only place robots can be used is in outer space, where there is no competition with human beings. And this, indeed, is the sort of situation that could conceivably take place. No matter how much people like myself (the cockeyed optimist) may think that a computerized society will be beautiful, we may come up against an absolutely immovable object, the suspicion of the average human being of being replaced by a computer. And in that case, it may be that a computerized society is not going to develop. I must say, though, that human beings are really not afraid of the computer. They may have already lost the fight, because computerization has already taken over society to such an extent that if every computer on Earth were suddenly to disappear, no industry of any size could probably continue for very long.

As an example, Doubleday and Company decided to switch computer systems, but they didn't do anything as dull as build up a new system, run it in tandem with the old one until they were quite satisfied, and then pull out the old one. No, they did it computer fashion. They pulled out the old system first, then they built up the second system, and are now engaged in the interesting process of trying to make it work. The result is that I cannot determine how many of my next novels have been distributed to the bookstores, nor do they know when they need to send out more copies. Who knows, they may ruin my entire theory because they have been getting along without a computer for a few weeks.

So if all computers were to disappear, not only would industries come to a halt, but the United States would no longer be able to collect income taxes (except what we voluntarily send), the Army wouldn't be able to do its work, and the space exploration industry would come to a halt. In short, we are already inextricably tied to the computer, and it is going to be difficult to stop it at any particular level.

Question: About 20 years ago, your view of the future was that humanity was basically in a disaster situation and might not survive the next 50 years. Furthermore, if we did survive the next 50 years, then getting through the next 50 years would be almost impossible. Have you changed your views or do you think we still have a difficult 30 years ahead?

Answer: I still think we are headed for a difficult 30 years. I mean, everyone here understands that someone could press the button tomorrow. We still face the possibility of a nuclear war. Little things happen that exacerbate feelings and make it difficult to talk sensibly. So we are constantly facing the destruction of civilization, not only by the instant blast of nuclear war but also by the continuing processes of increasing population, increasing pollution, foolish misuse of resources, and chemical damage (for instance, acid rain). We are in the uncomfortable position of facing the possible destruction of civilization either very quickly or a little more slowly in any of at least a dozen different directions. But, so far, we are still concentrating on localisms. Every nation worries about its neighbor, every nation sees as its primary concern the evil machinations of someone just across the border. We are so much more concerned about what the Soviet Union is doing than about what acid rain is doing, and the Soviet Union is more concerned about us than about any long-range danger to human beings. As long as that is true, by the time we wake up to the true dangers that face us all, if we ever do, it may be too late. So, yes, I still worry about civilization being a short-term process. Naturally, in a talk like this, I pretend that we will be wise enough to overcome the problems. As a matter of fact, in my presentation on space exploration, I discuss what I think is the only practical way that I can see of overcoming all this suspicion. It is a very slim chance indeed, but I don't see any other alternative.

Question: If computers eventually get rid of all the drudgery, will humans actually be capable of taking advantage of that, and how?

Answer: The easy answer is that if computers do all the stupid things we shouldn't be doing, then we will have time to do all the things we love. I mean, if you actually get fun out of alphabetizing cards, then do it. I do! I make all my own indexes, even though my publisher begs me to allow an expert to do them. He won't listen to me when I tell him that I am better than an expert. He says, "But your time is more valuable." I answer, "But I love it." I make all the little cards and I alphabetize them; I spend all evening long alphabetizing them, and I love it. If they ever design a computer to do it, I won't let it. So, we can keep our fun; what I am talking about is computers doing things that we don't want to do so that we can engage in creative endeavors. People will be designing, programming, and maintaining computers, working in scientific research or the arts, writing history books and novels, sculpting, or whatever it is that they want to do.

You respond, "Yes, but you are assuming that all human beings are creative in one way or another, and we all know that it isn't so because just look at all the noncreative people around you." To this I reply, "Well, you're looking at a ruined world." In medieval times, during the Dark Ages (at least in Western Europe), reading and writing were the province of a small group of clerks (clerics), and most other people, whether they were brutish aristocrats or bovine peasants, couldn't read or write. They didn't have any reason to read and write. If you had asked any of the few people who could read and write whether it was conceivable for either the aristocracy or the peasants to learn to read and write, the clerks would have said, "No, they are just animals, they are just brutes. Reading and writing are just for the very few with the kind of mind that we have." Yet, when the time came, we developed printing and the idea of mass education; it turned out that almost everybody could be taught to read and write, after a fashion. Reading and writing were not such unusual processes once we actually developed the educational procedures for them and made them economically feasible. In fact, once we developed a sufficiently complex culture, once we had a technological civilization, it became necessary for most people to read and write if they were to have any kind of job at all. And so it was possible.

We live in a world now in which education isn't geared for creativity and the kinds of jobs you have destroy any feeling of creativity you might have had to begin with. If you spend all day in the assembly line, what the heck are you going to develop? Your mind goes to pot. So it is amazing that the human brain, for the most part, is as well off as it is, that it hasn't totally dissolved into a kind of undifferentiated goo. I have great hopes that if you could get to youngsters, have them grow up with the aid of a computer, so they would have a one-to-one relationship with the wisdom of the world (the computerized wisdom of the world), they could follow up what interested them at their own speed and time. They would not be put to work doing things that are stupid enough for a computer to do but would be encouraged to do things that are more human. It may turn out that what we call creativity is a much more common feature of the human mind than we think. Maybe I'm lucky that I live in the last generation of the noncreative human society so that the rotten little stories I write stand out and manage to make a good living for me. Maybe if I lived in the 21st century, they would say, "Nice stories, Asimov, but anyone can do it."

Question: As computers get smarter, do you see us going through a scenario during the next 10 to 25 years like the one Toffler talked about in *The Third Wave?*

Answer: Actually, Toffler wrote a very plausible scenario. I find myself attracted to some of its features as a natural concomitant of such a computerization procedure. But, history and the human species have a way of surprising us. If I were to describe today what I thought life would be like, say, in the year 2050, and then bury it in the ground, when people dug it up in 2050 and read it, they would probably get a good laugh. Edgar Allan Poe had an amazing, first-rate imagination. He wrote a story the year he died about travel a thousand years later, in the year 2848. Do you know what he envisioned? He predicted balloons that could race across the Atlantic Ocean at a hundred miles an hour! If he could have been told that we were going to have objects that moved 10 times as fast and carried many more people in one hundred years, not one thousand years, he would have been astonished! He wouldn't have believed it! We always tend to underestimate the extent of human progress and what we can do if we don't destroy ourselves. It's very likely that what will be staring us in the face in the future will not be anything that anyone is predicting now, or if anyone is predicting it, no one else is paying any attention.

Question: What will your next works be?

Answer: On my 45th anniversary, my new novel, *Robots of Dawn,* is coming out. It's the third largest Elijah Baley novel, following *Caves of Steel* and *The Naked Sun*, and is a terrific book. Doubleday has printed 150 000 copies, which means they have confidence in it, but they still don't know how many are in the book stores right now.

I am also working on a revision to *Asimov's Guide to Science*, which is in its third edition, but it is now 12 years old, so you can imagine how out of date it is. I'm also trying to finish *Opus 300*, in time to have it published as my 300th book.

Our Future in the Cosmos—Space

Isaac Asimov

Our Future in the Cosmos—Space

Throughout the history of humanity, we have
been extending our range until it is now planet-wide,
covering all parts of Earth's surface and reaching to
the bottom of the ocean, to the top of the atmosphere,
and beyond it to the Moon. We will flourish only
as long as we continue to extend that range, and
although the potential range is not infinite, it is
incredibly vast even by present standards. We will
eventually extend our range to cover the whole of the
solar system, and then we will head outward to the
stars.

It frequently happened in my business as a writer, especially in
my younger days when I knew some pretty overwhelming editors,
that an editor would say to me, "I have a great idea for a story."
He'd slap me on the back and say, "Now go home and write it."
I would always think how easy it was for him to give me an idea
for a story, but it was I, not the editor, who had to sit down and
look at the most terrifying of all things: a blank page. In the
same way, it's fun to be introduced and have someone tell a lot
of exaggerations about me; however, then he sits down and I'm
the one who has to face the audience. I must say that it helps a
great deal to face an obviously friendly and intelligent audience. I
have brought almost the entire MENSA organization of this region
to this presentation, and, naturally, they take it personally when
I talk about intelligence. I am the international president of that
organization, not because of anything I have done but because of
a whim of the organization.

I want to discuss our future in the cosmos. One of the things I
think will mean the most to us and will make the future different
from the past is the coming of a "space-centered society." We are
going to expand into space, and I think it is fitting and right that
we should do so. All through the 50 000 years of *Homo sapiens*,
to say nothing of their hominoid precursors, humanity has been
expanding its range of habitation. We don't know exactly where
the first *Homo sapiens* made their appearance, but they have been

expanding until they now inhabit the entire face of the Earth. For the first time in human history, we are faced with a situation in which we literally have no place on Earth to expand. We have crossed all the mountains; we have penetrated all the oceans. We have plumbed the atmosphere to its height and the oceans to their depths. Unless we are willing to settle down into a world that is our prison, we must be ready to move beyond Earth, and I think we are ready. We have the technological capacity to do so; all that we need is the will. I think it is quite possible, starting now, to build settlements in space, to build worlds miniature in comparison to the Earth but large in comparison to anything we have done so far. These worlds, in orbit around the Earth, would be capable of holding tens of thousands of human beings.

This idea of space settlement seems odd to people; it doesn't seem inviting. When I suggested such an idea in an article I wrote a few years ago, I received a number of letters arguing against the possibility of space settlements. The arguments weren't based on economics; the main argument was that nobody would want to live in space. Nobody would want to leave his comfortable home on Earth. As nearly as I could tell from their addresses, all the people who wrote to me were Americans, and I presume that they knew American history. Americans should understand exactly what it means to leave their comfortable homes and to go to a completely strange world. This country was a wilderness at the beginning, and even after it was settled, it was a foreign land for most people. We in the United States are the descendents (unless any of you happen to be American Indians) of people who came here from other continents in search of something. Our forefathers, who came, at first, under harsh conditions, knew it would take them weeks to cross the ocean. They knew that if they met a serious storm, they would probably not survive. They also knew that when they landed, they would find a wilderness and possibly hostile natives. Yet, they still came. Between 1607 and 1617, 11 000 Englishmen came to the new colony of Virginia. In 1617, the population of Virginia was 1000. How was it possible for 11 000 people to come and yet to have only a population of 1000? The answer is easy; 10 000 died. Yet people continued to come. Why? They came because life in Europe, for many, was intolerable and because they wanted to come to a new land to start a new life. Whatever the risks, whatever the chances, if they succeeded it would be something new. It is this same desire that will drive people into space and cause them to populate as many space settlements as they can build. The chances of survival in space will probably be greater than those of the first immigrants to the colony of Virginia.

In their letters to me, some individuals wrote that people would not be able to endure the kind of engineered environment that would exist in the space settlements and that they wouldn't be able to bear not living close to nature as they do on Earth. Who lives close to nature here on Earth? There are millions of people on Earth who are never close to nature. I know; I live in the middle of Manhattan. I admit, I can look out the window and see Central Park from a distance, but I don't venture into it often. I think people should remember that the space settlements will probably be engineered to accommodate the comforts of the Earth's inhabitants. It is possible that people will be closer to nature in these settlements than in many places on the Earth today. People also wrote that the existence of space settlements would be unfair to the wretched of the Earth because the educated people would go into space and leave the less advanced people behind. That is probably precisely the reverse of what might happen. If we use the United States as an example, which classes of people came to this country? Obviously, the European ruling classes did not come; they were comfortable where they were. Why should they have left their homelands? The people who came to the United States were precisely those who hoped for something better, even if it meant a great deal of risk. Think of the passage engraved on the base of the Statue of Liberty: "Give me your tired, your poor, Your huddled masses yearning to breathe free, The wretched refuse of your teeming shore." I know those lines, you see, because in 1923, I was one of the "wretched refuse" who passed through Ellis Island. I've never forgotten 1923 because it was the last year in which people could enter this country without question. After that, the word went through the hallowed halls of Congress, "Asimov is in ... close the golden door." In 1924, the first strict quotas were placed on immigration. If I had tried to come a year later, I might not have been allowed to enter.

I imagine that when the time comes to begin emigrating to the space settlements, it will be hard work to make sure that not only the wretched of the Earth but also the educated people with usable skills are included. It's going to be just the reverse of what people are afraid of. In fact, I have also been told in some letters that space colonization would be unfair because only those nations with a heritage of rocket travel, space flight, or of high technology would be able to take advantage of this new frontier, leaving the rest behind. Again, that idea flies in the face of historical fact. As an example, when my father decided to come to the United States, he hadn't the slightest idea of what the ocean looked like; he had never seen it. He had no heritage of ocean travel. I don't think he had any idea what a ship looked like unless he had seen a picture of one, and even when he was on the ship, he didn't know what kept it afloat or how anyone on the ship could tell where they were going when they were in the middle of the ocean. I'm not sure I know, frankly. Yet he managed to get to the United States without any tradition or knowledge of seafaring because he had something else. I will tell you what people will need to get to a space settlement: it isn't a background in rocketry, it isn't technological know-how, it isn't any tradition of high technology. I'll tell you what it is if you will pay close attention because it's rather subtle. What they will need is a ticket, because someone else is going to take them.

Of course, you might ask yourself what these settlements in space will do for us. Will we settle in space just to make Asimov happy? Is there any other purpose to it? Yes, there is, because we're going to do a great many things in space that we can't do on Earth. For instance, 10 years ago, there was an energy crisis that most of us, perhaps, have now forgotten. These days we hear about an oil glut instead. Well the oil glut exists only because there was a world recession; there still is a recession, as a matter of fact. If we recover economically, the demand for oil will increase, the glut will disappear overnight, and OPEC will raise its prices again. There is a limited amount of oil and coal in the Earth (a great deal more coal than oil), but we could make do with coal for centuries except that it is increasingly dangerous to use. Coal is difficult to dig out and transport, and burning it results in air pollution, produces sulfur and nitrogen oxides that dissolve in the atmosphere's moisture to produce the acid rain that is destroying life in our ponds and lakes and is killing our forests. But quite apart from all this, if we continue to burn coal indefinitely, we will increase that fraction of the atmosphere which is made up of carbon dioxide. At the beginning of this century, approximately 0.03 percent of the air was carbon dioxide. This amount has increased almost 50 percent since then, and it will probably double within another half century. There won't be enough carbon dioxide in the air to interfere with breathing, but it may produce what we call "the greenhouse effect" because it tends to be opaque to infrared radiation. Ordinary sunlight that shines on the Earth passes through the atmosphere with little absorption and hits the Earth's surface. At night, the Earth reradiates a portion of this energy as heat (infrared radiation). If the level of carbon dioxide increases even slightly, this infrared radiation will have more difficulty getting out. It will be absorbed by the carbon dioxide, thus heating the atmosphere and raising the temperature of the Earth very slightly. It won't take much heating to cause the polar ice caps to melt, thus changing the climate of the Earth, undoubtedly for the worse! If you think that nuclear energy has the potential to make the Earth unlivable, so has the indefinite burning of coal and oil.

We are going to have to find some other sources of energy, and the only two sources of energy that will last as long as the Earth does are fusion energy and solar energy. I don't mean that we are going to have to depend solely on one or the other; there are other sources of energy that can be developed as well. There is geothermal energy, energy from under the Earth. There is biomass energy, the energy of the plant world. There is the energy of tides, wind, waves, and running water. All these can and will be used, but they are all relatively limited and there is no likelihood that they will supply all the energy we need. So, in addition to all these sources, we will need forms of energy that we can rely on in huge quantities forever. That brings us back to fusion energy and solar energy. We don't have fusion energy yet, although we've been working towards it for more than 30 years. We're not sure exactly what difficulties might exist between demonstrating it in the laboratory and developing huge power plants that will supply the world. We do have solar energy, but it's difficult to get in large quantities because it is spread thinly over the world. If we could get millions of photovoltaic cells (a kind of silicon cell that sets up a small electric current when exposed to light) and stretch them over half of Arizona (I only mention Arizona because there is usually a lot of sunshine there), we could perhaps supply enough energy for America's needs. If we did that in other parts of the world as well, we could supply the entire world. There is no doubt, however, that setting up solar cells (photovoltaic cells) on the Earth's surface is not very efficient. For one thing, there is no solar energy for the cells to absorb during the night. Even in the daytime under the best conditions (for example, in a desert area without fog, mist, or clouds), clear air absorbs a substantial portion of the sunlight, especially if the Sun is near the horizon. And of course, you also have the problem of maintaining these cells against nature's effects and against vandalism.

For these reasons it might be more reasonable to build a solar power station in space. Under such conditions, we could make use of the entire range of solar energy 98 percent of the time, because the stations could easily be positioned so they would fall into the Earth's shadow only 2 percent of the time, at the equinoxes. A chain of these stations around the Earth would allow most of them to be in the sunshine all the time. Optimists have calculated that in space, a given area of solar cells will provide 60 times more energy than on the Earth's surface. We can then imagine this chain of power stations circling the Earth in the equatorial plane at a height of approximately 22 000 miles above the Earth's surface. At this distance their orbital position will just keep time with the surface of the Earth as it rotates about its axes. If you stood on a spot at the equator and looked up at the sky with a sufficiently strong telescope, you could see the solar power station apparently motionless above you. I feel a certain proprietorship toward this idea of a space station. It was advanced about 20 years ago by people at the AVCO Corporation in Massachusetts, but about 40 years ago I wrote a story called "Reason" in which I talked about just such a power station. Of course, I missed the important point of having it in orbit around the Earth. I described it in an orbit similar to Mercury's around the Sun so that it could get even more energy. I ignored the fact that it would be awfully difficult to aim it at Earth from such a distance; in science fiction stories, you can dismiss such problems by saying that an advanced technology won't find it difficult to achieve. Nevertheless, solar power stations are my idea, and I'm proud of it!

There are a great many other things we could do in space. We could set up mining stations on the Moon and have laboratories in space to perform experiments you wouldn't want to do on Earth because of the risks involved to the population. Some years ago, people were very worried about recombinant DNA research. They feared that scientists would come up with a new strain of bacteria which would get out into the biosphere, and once it did, you would never get rid of it. It was like Pandora's box, when she opened it, all the ills of the world flew out and have plagued humanity ever since. In this same vein, suppose that for some very good reason, from the standpoint of research, scientists developed a strain of *E. coli* (a common bacteria that lives in the human large intestine) which had a very interesting chemical property that they wanted to study. But at the same time, it might turn out that this strain would make people prone to diarrhea. Suppose this strain is released to the world. People always speak about the danger of developing a "black death" germ that would kill everybody it touches and how terrible it would be if it were released. I don't think we have to be that extreme. An *E. coli* strain that would bring about diarrhea could be extremely disturbing to the entire Earth.

However, at the time when people spoke and worried about recombinant DNA research and worked up all kinds of horrible nightmares in connection with it, I believed it might turn out to be important and valuable research. It occurred to me then that this research might develop strains of bacteria that could form insulin, other hormones, and certain blood fractions, things that we need in quantity and can't get in the usual way. Recombinant DNA research might produce microorganisms that could fix nitrogen from the atmosphere and form terrific fertilizers or other microorganisms that could consume hydrocarbons under certain conditions and clean up oil spills. The research might simply give us information about the organization of living cells so that we could better understand what causes and what might cure cancer, or arthritis, or any of the other degenerative diseases that are now the major inflictions of the human race. How nice it would be to set up a space laboratory in Earth orbit in which the recombinant DNA research could be done. It wouldn't matter how dangerous the research was. I suppose it would still be mathematically possible for bacteria to escape and infect the Earth, but the chances would be far less than if the work were actually done on Earth. We could perform many such dangerous experiments in space. We could establish fission and fusion power stations in orbit and not have to worry about Three Mile Island incidents. Naturally, people working in the stations would still be exposed to these dangers, but they would be relatively few in number. They would be volunteers and specialists, and would know the risks involved. That is a different matter than doing research surrounded by millions of innocent people who are not aware of the risks.

We can also build observatories in space. I always said that we should set up a telescope in space which could look at the universe from outside the Earth's atmosphere, and now events are finally catching up with my imaginings. Even at its best, the atmosphere obscures. It's warm and its temperature varies so that there are always shifting columns of air. Whenever you look at the sky, it is like looking at it through frosted glass or through something that is transparent but trembling. If you have ever watched a television screen that for some reason is shaking, you realize how annoying it can be. When an astronomer looks at the heavens, the image is always shaking. That's why stars twinkle and why you can't see Mars' surface from Earth any clearer with a large telescope than with a small one. The large telescope shows you a larger Mars; it also shows you larger twinkles, which obscure the surface. If we could get outside the atmosphere, we could see much more clearly. There would be no twinkles because the vacuum doesn't interfere with viewing like the atmosphere does. We would be able to see the distant galaxies in great detail and possibly tell more about the beginning and the end of the universe. We could see all kinds of unusual stars in greater detail and learn more about stellar evolution and about some of the queer beasts in the astronomical zoo. But I always said this entirely on faith, and sometimes I wondered to myself, "What if we put a telescope out there and it doesn't find anything!" Well, those are the breaks of the game, but I would have been very disappointed.

Recently the United States launched the IRAS (infrared astronomy satellite) to examine the universe in the infrared range. It saw a great deal that we can't see from the surface because our atmosphere absorbs infrared radiation. One of the things the telescope looked at was the star Vega. It turns out, this star emits a surprising quantity of infrared radiation. However, astronomers looked more closely at this phenomenon and determined that the infrared radiation was coming not from the star itself, but from an annular region all around it. Apparently, there are colder objects circling Vega which absorb some of Vega's light and emit it as infrared radiation. These objects are not simply a shell of dust around Vega; they are larger particles, and the implication is that they are in the process of condensing into a planetary system. This is the first time we have ever acquired observational information concerning the development of any planetary system other than our own. There are various theories concerning the formation of planetary systems, and if these theories are correct, then almost every star should have a planetary system. For obvious reasons, we have not been able to actually see the planets of the distant stars. Stars are very far away and any planet shining only by reflected light can't reflect enough light to show up in our telescopes. Even if they did, they are so close to the star that their light would be drowned out by the much brighter light of the star. But now, as a result of IRAS, we have seen what seems to be a planetary system in the process of formation about another star, which makes us feel a little more confident about our theories of the way planetary systems should form. We now feel a little more confident about saying that stars have planets, as a general rule. Why does this star theory matter to us on Earth? There is a long chain of reasoning; there are many stars in the universe and a certain percentage of them resemble our Sun. If all the stars have planetary systems and these Sun-like stars have planetary systems, then a certain percentage of these planets ought to be Earth-like. If Earth-like planets exist, then they probably have developed life, and if there are this many life-bearing planets, one of them should develop intelligent life. Perhaps one of these has developed a technological civilization that we can detect or, perhaps, they are trying to contact us. This chain of reasoning causes some astronomers to feel certain that the universe has a great many technological civilizations, of which we are only one. However, this chain is so attenuated, so weak, and so highly theoretical that it is perfectly possible to argue, as some astronomers do, that the chain is broken at one or more points and that we may be the only technological civilization in our entire galaxy. It would be nice to know the answer. A telescope in space has already given us some reason to think that there may be other

87

technological civilizations in space besides our own. Who knows what else such instruments may discover?

Another kind of structure in outer space is factories. There is no reason why a good proportion of our industrial factories couldn't be placed into orbit. Space has very unusual properties that may be helpful to us. It has unlimited vacuum, zero gravity, the possibility of high and low temperatures, and hard radiation. There are a great many things we can do in space that we can do only with difficulty, if at all, on Earth. Most important of all, when we have a factory in space, any unavoidable pollution that it produces can be discharged into space.

Space is huge compared to the surface of the Earth. Some people argue that to earlier generations the ocean seemed huge and capable of absorbing any amount of pollution. But now we are in danger of poisoning the entire atmosphere. Some people argue that in the future we may be so casual about releasing pollutants into space that we may gradually poison all the space around ourselves. However, that won't happen, for not only is space literally millions of times more voluminous than the biosphere and not occupied by trillions of living things, but it is also true that nothing we release into space is going to stay there because of something called the solar wind. The Sun emits speeding particles in every direction; it has been doing this as long as it has been in existence and will continue to do this for billions of years. This solar wind will push the pollutants out beyond the orbit of Mars, beyond the asteroids and into the outer solar system, where there is a trillion times more room than in the Earth's neighborhood. The solar wind has a natural ventilating effect. This is important because it means that perhaps Earth can get rid of its "dark satanic mills" (to quote William Blake, who wrote in the first decades of the 19th century) without abandoning industrialization. People who view industrialization as a source of the Earth's troubles, its pollution, and the desecration of its surface, can only advocate that we give it up. This is something that we can't do; we have the tiger by the tail. We have 4.5 billion people on Earth. We can't support that many unless we're industrialized and technologically advanced. So, the idea is not to get rid of industrialization but to move it somewhere else. If we can move it a few thousand miles into space, we still have it, but not on Earth. Earth can then become a world of parks, farms, and wilderness without giving up the benefits of industrialization.

All this will be possible because we will have structures built in space. Who will build these space structures? It seems to me that it's an unnecessary expense to have them built by commuters. It wouldn't make sense to send people into space every morning and have them come back every evening or, even, to send them up every spring and have them come back every fall. We would want the people who are busy constructing the necessary structures in space, maintaining them, and improving them to be people who live in space. Why should the people of the space settlements labor to do this? They would share in the benefits to be derived from it, and, I suppose in the last analysis, they would do it for money. In other words, in exchange for their labor, they would get some things that would otherwise exist only on Earth. There would be a fine economic balance that I will allow economists to work out. The fact of the matter is that we would have a much larger, more variegated, and versatile world; it would be much richer and more advanced in knowledge so that we would look back on the present and think of it as a dark age when human beings lived only on Earth.

The space settlers, who will live on these worlds in orbit, will be the cutting edge of humanity for the future. These are the people who will move farther out into the solar system. It was difficult to reach the Moon although the flight took only 3 days. Imagine the problems for us to reach Mars when it might take months of travel or to reach the outer solar system when it might take years of travel? We are not really built for space flight; we are used to living on the outside of a huge world, not in the inside of a spaceship. We are used to a system of cycling air, food, and water that is so large that we are unaware of the actual process. We don't know where the pure sparkling water that we drink comes from, and we don't care. We don't know how the plants that we eat grow or what they use for food, and we don't care. We don't know what processes the atmosphere uses to clean itself. But if we lived in a spaceship, we'd know. We'd know that our air was manufactured from the carbon dioxide that we exhaled and that the food and water were once part of our waste products. (That's also true on Earth, of course, but we're not aware of it.) We would also be subjected to gravitational systems that would not be like those on Earth; they would vary. For all these reasons, space flight seems unnatural to us. But to the space settlers, who would arrive by space flight and live and work in larger versions of a spaceship, these conditions would seem natural. They might run mines on the Moon, and they would travel in a spaceship that would be very much like the space stations in which they would live (maybe a little smaller but that's all). They would be living inside a world with tight cycling and varying gravitational forces. They would be the natural pioneers. They, not we, would be the Vikings, the Phoenicians, the Polynesians of the future. They would make the long trips to Mars and the asteroids and learn how to mine the asteroids. They could travel out into the solar system and make plans to reach the stars someday. All we can do here on Earth, maybe, is reach the Moon. From worlds in orbit around the Earth, we can reach all the rest.

Beyond all these material things that space exploration can bring us, there is something completely immaterial that counts more than anything else. One thing that can stop us from going into space, from realizing what I consider a glorious possible future for humanity, is the fact that here on Earth, most people, especially those in power, are far more concerned with the immediate threat from other countries than they are with the possible dangers to civilization as a whole. How much of any country's mental energy, money, effort, and their emotion is directed towards saving civilization from destruction by pollution, overpopulation, or war, and how much is spent maintaining armed forces because of the danger from neighboring countries? You know the answer; the world is now spending 500 billion dollars every year for war and preparations for war. That's half a trillion dollars every year spent on forces that we don't dare use, or if we do use them, it is only to wreak destruction. The United States and the Soviet Union quarrel over differences that may be extremely important, but if the quarrel extends to the point of a nuclear war that destroys civilization, the differences become inconsequential.

How are we to prevent this whole thing from happening? There is one example in history that is very unusual. From 1861 to 1865, the United States fought the War Between the States, and many of its most epic battles were fought on Virginia's soil. One side lost; one side won. For a period of years, the winners showed no mercy as far as the losers were concerned, and the losers lived under occupation forces. The South has lived with this loss ever since, and yet the bitterness passed. This is not to say that the South has forgotten the Confederacy (of course it hasn't), but it's not forever laying plans to reestablish it. It hasn't maintained an attitude of unforgiveness; it doesn't say, "We will never forget." It doesn't always try to find allies abroad to help it reestablish itself. We have reunited into a single nation. How did we manage to do that, when there are other places on Earth in which the mutual hatred has continued undiminished because of things that happened thousands of years ago, and people refuse to forget? My theory is that after the Civil War there was a period of the development in the West, in which the North and the South could take part indiscriminately. People from both sides traveled westward and established the new states, and in the positive task of developing the western half of the United States, the old quarrels were forgotten. What was needed was something new, something great, something growing into which the old problems would sink into insignificance. It was just our good fortune that we had the development of the West to occupy our minds in the half century after the Civil War.

I have a feeling that if we really expanded into space with all our might and made it a global project, this would be the equivalent of the winning of the West. It's not just a matter of idealism or preaching brotherhood. If we can build power stations in space that will supply all the energy the world needs, then the rest of the world will want that energy too. The only way that each country will be able to get that energy will be to make sure these stations are maintained. It won't be easy to build and maintain them; it will be quite expensive and time-consuming. But if the whole world wants energy and if the price is world cooperation, then I think people are going to do it.

We already cooperate on things that the whole world needs. International organizations monitor the world's weather and pollution and deal with things like the oceans and with Antarctica. Perhaps if we see that it is to our advantage to cooperate, then only the real maniacs will avoid cooperating and they will be left out in the cold when the undoubted benefits come in. I think that, although we as nations will retain our suspicions and mutual hatreds, we will find it to our advantage to cooperate in developing space. In doing so, we will be able to adopt a "globalist" view of our situation. The internal strife between Earthlings, the little quarrels over this or that patch of the Earth, and the magnified memories of past injustices will diminish before the much greater task of developing a new, much larger world. I think that the development of space is the great positive project that will force cooperation, a new outlook that may bring peace to the Earth, and a kind of federalized world government. In such a government, each region will be concerned with those matters that concern itself alone, but the entire world would act as a unit on matters that affect the entire world. Only in such a way will we be able to survive and to avoid the kind of wars that will either gradually destroy our civilization or develop into a war that will suddenly destroy it. There are so many benefits to be derived from space exploration and exploitation; why not take what seems to me the only chance of escaping what is otherwise the sure destruction of all that humanity has struggled to achieve for 50 000 years? That is one of the reasons, by the way, that I have come from New York to Hampton despite the fact that I have a hatred of traveling and I faced 8 hours on the train with a great deal of fear and trembling. It was not only The College of William and Mary that invited me, but NASA as well, and it is difficult for me to resist NASA, knowing full well that it symbolizes what I believe in too.

Question: The first book of yours that I read was *I, Robot.* In your opinion, how close are we today to the world you described in that book?

Answer: Although the book was written in 1939, those robots were very intelligent and human-like in their capacity. As yet, the robots we use today are merely computerized arms that can do one specialized job. So, we're not very close, but we're heading in the right direction. Although I have never done any work on robots and know almost nothing about the nuts and bolts, I think that I came close enough that I am almost the patron saint of robotics. Most of the people who work in robotics obtained at least some of their early interest in the field by reading my books. I was the first person to use the word robotics, and I spoke of the *Handbook of Robotics*, from which I quoted my three laws. I said they were from the 56th edition, in 2058 A.D. Now someone is actually in the process of putting out the first edition of that book, and they've asked me to write the introduction. I guess the people who are working in robotics see themselves moving toward the world I described 40 years ago, and I'm willing to accept their judgment.

Question: Why do you restrict yourself to looking for Earth-like planets in the search for technological civilizations, why not Jupiter-like planets, for instance, or Pluto-like planets?

Answer: If we assume that there can be life even under widely varying conditions, we make the problem perhaps a little too easy. There is also the chance that life evolving under such conditions might be so different from human life in very basic ways that we will not be able to detect it or to understand that it is a technological civilization even if we encounter it. As our information and knowledge grow, we might be able to widen our view to recognize life and civilization of widely different kinds. But to start with, acknowledging our own limitations and the fact that we know so little, we are looking for technological civilizations sufficiently like our own to be perhaps recognizable. So at the start, but not necessarily forever, we restrict ourselves to Earth-like planets.

Question: Do you think, because our bodies are fragile and we have limited life spans, that what we now know as humanity would ever be replaced by inorganic intelligence?

Answer: I believe that computers have a kind of intelligence which is extremely different from our own. The computer can do things that we are particularly ill adapted to do. Humans don't handle rapid intricate calculations very well, and it's good to have computers do them. On the other hand, we have the capacity for insight, intuition, fantasy, imagination, and creativity, which we can't program into our computers, and it is perhaps not even advisable to try because we do it so well ourselves. I visualize a future in which we will have both kinds of intelligence working in cooperation, in a symbiotic relationship, moving forward faster than either could separately. The fact that we are so fragile and short lived is an advantage in my way of thinking. In *Robots of Dawn*, I compare two civilizations; one is our own in which people are short lived, and the other is that of our descendants in which they are long lived. I point out the disadvantage to the species as a whole of being long lived. I won't repeat the arguments, because if I don't you may storm the bookstores out of sheer curiosity to see what I've said.

Question: One of the great themes of science fiction is the settlement of other planets. Is there any place in this solar system or nearby that might be habitable?

Answer: As far as we know, there is no world in our solar system that is habitable by human beings without some form of artificial help. The Moon and Mars, which come the closest to being tolerable, will require us to build underground cities or dome cities, and if we venture on the surface, we will have to wear space suits. This is not to say that it will not be possible someday to terraform such worlds and to make them habitable; but I honestly don't know if it will be worth it for us to do so. As to planets circling other stars, we do not really know of such planets in detail. We suspect their existence, and we figure statistically that a certain number of them ought to be habitable, but we have yet to observe any evidence of such a thing. It is still very much in the realm of speculation.

Question: You made the analogy between the migration from Europe at the turn of the century and possible future migrations to space stations and other planets. It has been shown that as a result of our technology, people in this country are taller, heavier, better built, and able to set new records in endurance and physical capabilities. Would you speculate about the effect that living in space stations might have on the human body and its evolutionary potential?

Answer: It is hard to tell. I suspect that people will make the environment of these space settlements as close to that of Earth as possible. But in one respect, they will have problems; there is no way that they can imitate Earth's gravitational field. They can produce a substitute by making the space settlement rotate, so that the centrifugal effect will force you against the inner surface and mimic the effects of gravity. But it won't be a perfect imitation; there won't be a Coriolis effect and, also as you approach the axes of rotation, the gravitational effect will become weaker. The people who will live in a space settlement will be exposed to variations in the gravitational effect far greater than any you can possibly feel on the surface of the Earth. This may give rise to all sorts of physiological changes in human beings. I don't know what they will be; we can't know until we actually try living in space. So far, people have been subjected to essentially zero gravity for as long as 7 months at a time without apparently permanent ill effects. But human beings have never been born at zero gravity or under varying gravitational conditions; they have never developed and grown up under such conditions, and we can't be sure what the effects will be. From an optimistic standpoint, I suppose that under such conditions human beings will develop a greater tolerance of gravitational effects than they now possess. This will further prepare them for life in the universe, whereas we ourselves have been specifically evolved and conditioned for life in one very specialized place in the galaxy. The overall effect may be to strengthen the human species; at least, I'd like to think so. The future will tell us if that is so.

Question: In your opinion, when will there be solar power stations in orbit and manned ventures to Mars, considering the technological leaps with the Space Shuttle and the Soviet's Salyut space stations?

Answer: It is hard to say when solar power stations in space will be developed. It's up to the human governments that control the money and the manpower. If we begin to cooperate and make a wholesale attempt, we could have solar power stations in space before the 21st century was very old. In other words, someone as young as the person who asked me this question, may see space stations by the time he is middle-aged. But on the other hand, if we choose not to do it, then we may never have these stations in space. The choice is ours. We can choose to develop space or we can choose world destruction. I'm at a loss to state in words how desirable the first alternative is and how likely the second alternative is.

Question: What kind of timetable do you envision for humanity's exploration of space, and what good or harm do you think is done by prospace groups?

Answer: Well, we can't expect things to happen too quickly.

95

The region that we now call the United States was being settled for nearly two centuries before this country came into existence. We've celebrated our bicentennial as a nation, but in a little over 20 years we're going to have to celebrate the tetracentennial of our existence as a community on American soil, from the establishment of Jamestown in 1607. If it took nearly two centuries to settle the United States to nationhood, it might take that long to establish a space community strong enough to be independent of the Earth. On the other hand, things move more quickly now; we're more advanced. It may take less than a century to do so if we really try hard. As for the effects that prospace organizations might have, I'm not a sociologist so I just don't know. I'm in favor of prospace organizations doing their best to persuade human beings to support space exploration. I don't know how that can be bad.

Question: Assuming that we do not annihilate ourselves, what is your view of how life on Earth will evolve, both humans and other life forms?

Answer: You must understand that evolution naturally is a very slow process and human beings can well live for 100 000 years without many serious changes. On the other hand, we are now developing methods of genetic engineering which will, perhaps, be able to wipe out certain inborn diseases, or correct them and improve various aspects of the human condition. I don't know how we will develop or what we will choose to do; it's impossible to predict.

Question: How long do you think it will be before people live in outer space?

Answer: That's entirely up to us. In a way, we've had people living in outer space already, ever since the first Russian cosmonaut spent 1 1/2 hours in space. We have now had people living in outer space for 7 months at a time; in fact, one Soviet cosmonaut lived in outer space for 12 months over a period of 18 months. So we've had people living in outer space already, and I'm sure we'll have more and more of them for longer and longer periods of time.

☆ U.S. GOVERNMENT PRINTING OFFICE: 1985—470-563

www.ingramcontent.com/pod-product-compliance
Lightning Source LLC
Chambersburg PA
CBHW051339170526
45166CB00002B/875